T0326353

Combined Cooling, Heating and Power

Combined Cooling, Heating and Power

Decision-Making, Design and Optimization

Masood Ebrahimi
Assistant Professor at University of Kurdistan, Iran

Ali Keshavarz
Associate Professor at K.N. Toosi University
of Technology, Iran

ELSEVIER

AMSTERDAM • BOSTON • HEIDELBERG • LONDON • NEW YORK
OXFORD • PARIS • SAN DIEGO • SAN FRANCISCO • SINGAPORE
SYDNEY • TOKYO

Elsevier
Radarweg 29, PO Box 211, 1000 AE Amsterdam, Netherlands
The Boulevard, Langford Lane, Kidlington, Oxford OX5 1GB, UK
225 Wyman Street, Waltham, MA 02451, USA

Notice
No responsibility is assumed by the publisher for any injury and/or damage to persons
or property as a matter of products liability, negligence or otherwise, or from any use or
operation of any methods, products, instructions or ideas contained in the material herein.
Because of rapid advances in the medical sciences, in particular, independent verification
of diagnoses and drug dosages should be made

British Library Cataloguing in Publication Data
A catalogue record for this book is available from the British Library

Library of Congress Cataloging-in-Publication Data
A catalog record for this book is available from the Library of Congress

ISBN: 978-0-08-099985-2

For information on all Elsevier publications
visit our website at http://store.elsevier.com/

Dedication

There is no shortcut to writing a book.
It needs the absolute support of family.
To my wife and my little son

Masood Ebrahimi

To my family
By whom I was encouraged
To my students
By whom I was inspired

Ali Keshavarz

Contents

Preface

Investigation of combined cooling, heating, and power (CCHP) systems by different researchers in the last decade have revealed that CCHP is reliable and economical. It saves fuel and reduces air pollution and greenhouse gases. It is also safer with respect to centralized power generation systems in critical situations such as war, terrorist attacks, and natural disasters. Due to these essential characteristics it is predicted that the use of distributed power generation such as CHP and CCHP systems will develop rapidly in the near future.

This book is written as a guide for CCHP researchers, designers and operators. The contents of this book can accelerate the new research and save considerable time for those who are new to this topic. This book also presents some general guidelines for operation and maintenance of CCHP systems.

In Chapter 1 most of the published research from 2002 to 2013 is summarized. The main attention of this literature review is to present the main design and decision-making criteria that concern researchers and designers. In addition different CCHP cycles presented by different investigations are discussed and presented. Through this literature review, the reader will become familiar with many CCHP cycles and their components.

In Chapter 2 the main technologies used in the basic CCHP cycles are introduced. These technologies include different prime mover types such as steam turbines, gas turbines, reciprocating internal combustion engines, micro-gas turbines, micro-steam turbines, Stirling engines, fuel cells and thermal photovoltaic technology. The basic CCHP cycles that can be designed with different prime movers are also presented. Furthermore, cooling system technologies, especially those that are thermally operated, are presented in this chapter. Thermally activated cooling systems include absorption chillers, adsorption chillers, solid and liquid desiccant dehumidifiers, and ejector refrigeration systems.

In the third chapter the main recommended evaluation criteria for use in the decision-making and design steps are introduced and formulated. The criteria are classified into four main groups: technological, economical, environmental, and miscellaneous. Every criterion is also divided into several subcriteria. For example, the technological subcriteria include fuel saving, exergy efficiency, overall efficiency, operation in partial load, maturity of the technology, recoverable heat quality, user-friendliness of control and regulation, etc. The economical subcriteria include initial capital cost, operation and maintenance cost, net present value, payback period, internal rate of return, net cash flow, etc. The environmental subcriteria may include reduction of air pollution such as CO_2, CO, and NO_x as well as noise. The miscellaneous subcriteria may include many parameters such as the footprint, ease of maintenance, import and export limitations on CCHP components, lifetime, etc.

In the fourth chapter, two methods are presented for decision-making for certain CCHP components, such as the prime mover, cooling, or heating systems. Decisions are made based on the fuzzy logic, and the grey incidence approach. These methods are called multicriteria decision-making methods. In this chapter, as an example, the prime mover of a CCHP system is chosen among several options for various climates.

To design a CCHP system, load calculation is one of the most important steps. In Chapter 5 different load calculation methods that can be used to design a CCHP system are presented. In addition, the load calculators and websites that can be used for finding necessary weather information are introduced. Also in this chapter the energy demands of a sample building in five different climates are calculated and compared.

In Chapter 6 different design methods for CCHP systems are presented. These methods include the classic maximum rectangle method (MRM); developed MRM; energy management sizing methods including FEL, FTL, and FSL; the thermodynamical sizing method; the thermoeconomical sizing method; the multicriteria sizing function; and the fitness function method. In addition a CCHP system is designed for five sample climates by using different sizing methods and the results and advantages and disadvantages of sizing methods are compared as well. This chapter is especially helpful for designers.

Using renewable energy sources besides fossil fuels in CCHP systems increases the capabilities of this new technology. In the seventh chapter solar heat in particular is studied for use in CCHP systems. A solar collector is coupled with the CCHP cycle and a method is proposed to determine the optimum direction and size of the collector in five climates.

Since usually the surplus heat is wasted in heating systems and basic CCHP cycles in particular, storing the surplus heat for reuse at a later time can increase the advantages of the CCHP system. In Chapter 8 the principals of thermal energy storage (TES) systems are introduced and the CCHP cycle designed in the previous chapters is equipped with the TES system.

Operation and maintenance of CCHP systems is a great job. A proper operation and maintenance program maintains the CCHP cycle in optimum condition, increases its life span and reliability, and decreases maintenance costs. In Chapter 9 the basics of operation and maintenance, pre-commissioning, commissioning, post-commissioning, and troubleshooting of CCHP cycles are presented as a guideline for operators.

Finally in Chapter 10 the mutual benefits of CCHP cycles for consumers and the government are discussed; this positive impact highlights a bright future for CCHP systems.

We believe that CCHP systems can be reliable as an energy conversion equipment because they use fuel energy very efficiently in comparison with the conventional systems for producing cooling, heating, and power. They are also helpful in reducing the risk of global warming. They are safer in critical situations such as war, terrorist attack, and natural disasters such as flood, earthquake, and thunderstorms.

We believe that this book is just an introduction for the decision-making process for CCHP systems and the design, and optimization of them. We appreciate every proposal, correction, question, comment, or criticism on this book with our heart and soul. The readers can write letters about the book to Dr. Masood Ebrahimi, University Of Kurdistan, Kurdistan Province, Iran. In addition they can send email to ebrahimi_masood@yahoo.com or ma.ebrahimi@uok.ac.ir.

CCHP Literature

1.1 Introduction

Conventional separate production of cooling, heating, and power has been used since the first power plant was built in 1878.[1] In this book it is called SCHP (separate production of cooling, heating, and power). In SCHP, electricity is generated in conventional centralized big power plants that can produce hundreds of megawatts of electricity. Then it is transmitted and distributed through the electricity grid to feed the terminal consumers in different sectors, such as residential, industrial, commercial, etc. The heating and cooling demands of the terminal consumers are provided by the grid electricity, combustion of fossil fuels, and/or renewable energy resources such as the sun, geothermal resources, etc.

On the contrary, in the combined production of cooling, heating and power, which in this book it is called CCHP,[2] electricity is generated in the vicinity of the consumer, and the waste heat of the power generation unit can be recovered to produce heating or cooling when the need arises. Due to generation of electricity in the vicinity of the consumer, the transmission and distribution losses (T&D losses) are omitted.[3] The CCHP system increases the energy efficiency by recovering heat loss and omitting grid loss. A schematic of the SCHP and CCHP systems are presented in Figure 1.1. In the SCHP system, electricity is received from the grid. The overall efficiency of the power plant, transmission, and distribution grid is assumed to be 30%, heating is provided by a boiler with the effectiveness $\varepsilon = 80\%$, and cooling is provided by an absorption chiller with a coefficient of performance (COP) of 0.7. In the CCHP unit however, electricity is provided by the prime mover, heating is prepared by recovering the waste heat, and cooling is produced by using the recovered heat in the absorption chiller. The figure shows that to provide the same demands for a building the CCHP overall efficiency can be as much as 30% higher than the SCHP; in addition the CCHP consumes 37.76% less fuel than SCHP. This higher efficiency means considerable fuel savings and pollution reduction, leading to more economically efficient system.

CCHP systems have additional advantages. They are less fragile than centralized power plants when natural disasters such as earthquakes occur. Because they are distributed, they can run independently from the grid and also can be designed to run with different fuels. They are also less sensitive to terrorist attacks or war, because it is almost impossible to attack all of the CCHP systems. A study after the September 11th

---

[1]The world's first power plant was built and designed by Sigmund Schuckert in the Bavarian town of Ettal and went into operation in 1878. The station consisted of 24 dynamo electric generators, which were driven by a steam engine.

[2]In some research the CCHP concept is also called trigeneration.

[3]A major vendor of superconducting conductors claims that the HTS cable losses are only half a percent (0.5%) of the transmitted power compared to 5-8% for traditional power cables. Source: www.nema.org.

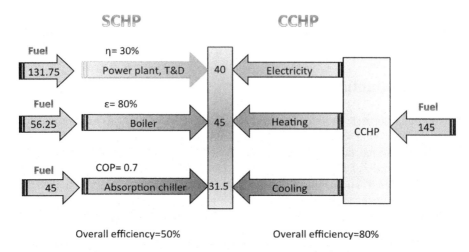

Figure 1.1 Comparison between SCHP and CCHP systems.

terrorist attacks suggested that a system based on distributed generation plants may be five times less sensitive to systematic attack than a centralized power system [8].

In addition, a CCHP system can make money if well designed and managed. A CCHP system can earn money by saving fuel and selling the surplus electricity to the grid. According to the research, a payback period of less than five years is achievable for systems that can work at least 15 years. This means the CCHP system produces significant economic benefits in the last 10 years of its life.

The advantages of CCHP systems make them a very good solution for those who care about energy resources, the Earth's environment, air pollution, greenhouse gases, and safety of power generation, transmission and distribution. To help in utilizing these advantages, this book is prepared as a guideline for the decision-making process for CCHP system and their design and optimization. In this chapter a literature review of CCHP systems highlighting the last decade will be presented. In this review we have focused on the different CCHP cycles and evaluation criteria used by researchers.

1.2   CCHP in the Last Decade

In order to be familiar with the advances of and recent research about CCHP systems, we have done a review of the published research in the last decade. In this literature review, we have focused on different CCHP cycles, which are presented numerically and/or experimentally. In addition, the design criteria and considerations for each CCHP cycle are also summarized. This review shows how the energy or waste energy is used in different CCHP technologies. It also draws a comprehensive picture of the technologies that are currently used or being actively researched for use in CCHP systems. This review starts in 2002 and will help researchers, designers, and students become familiar with the concept of CCHP very quickly and accelerate their learning.

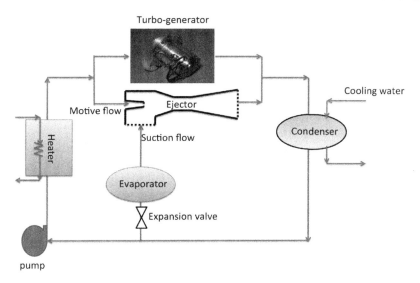

Figure 1.2 The CCHP cycle and the mini turbo-generator proposed by [1].

In 2002, Ref. [1] presented an empirical work involving an ejector heat pump coupled with a Rankine cycle that was proposed to meet the heating, cooling, and electricity demands of a residential building. The cycle was designed to use solar energy or natural gas as the input energy resource. In Figure 1.2 the cycle and the mini turbo-generator used in this experiment are presented.

In the cycle presented in Figure 1.2, the impact of the three working fluids n-pentane, water, and R134a on the coefficient of performance (COP) of the cooling cycle was investigated. Two prototypes with cooling capacities of 2 and 5 kW were built. The nominal output power of the turbo-generator was 1.5 kW and water was used for its cooling. One of the cycles was equipped with a 20 m^2 solar collector to be used in the UK, and the other one used natural gas as the fuel and operated in Portugal. The results show a mean COP of 0.3, and electrical efficiency of 3% to 4% in ambient temperature of 20°C. The boiler temperature in these experiments was 95°C, pump and fan consumption was negligible, and turbine efficiency due to frictional loss and water cooling was 28%. This research concludes that improvement of turbine technology has a significant impact on cycle performance. The cycle also shows improvement in CO_2 reduction in comparison with conventional systems, which consume coal as the fuel. This research also defined an economic criterion called *cost figure (CF)* as follows:

$$CF = \frac{capital\ cost + running\ cost + maintenance\ cost}{(energy\ delivered/year) \times number\ of\ years}\ (euro/kW\ h) \qquad (1\text{-}1)$$

The results showed that the economic criterion improves when the cycle is used for trigeneration rather than power and cooling generation only. They also concluded that the cycle is especially economical when the cooling load is high and the cycle is used to provide domestic hot water (DHW) as well. While Ref. [1] concluded that promotion of small- and micro-scale turbines is necessary to improve performance of CCHP systems, Ref. [2] in 2002 reported the characteristics of micro gas turbines (MGTs), and also presented the main producers of MGTs around the world. This reference also reports that among different applications of MGTs, the most popular use is in combined heating and power (CHP) systems, because it has a high-quality exhaust gas that makes it very suitable for district heating.

In 2003, Ref. [3] considered some hotels in the Euro zone and based on the hotels' demands designed the main components of a CCHP system including the prime mover and absorption chiller. The design criteria of the CCHP system that they considered include an energy utilization factor (EUF), artificial thermal efficiency (ATE), fuel energy savings ratio (FESR), and exergy efficiency (π). These criteria are defined as follows:

$$EUF = \frac{E_{PM} + H_{dem}}{F_{CCHP}} \qquad (1\text{-}2)$$

$$ATE = \frac{E_{PM}}{F_{CCHP} - \dfrac{H_{dem}}{\eta_b}} \qquad (1\text{-}3)$$

$$FESR = 1 - \frac{F_{CCHP}}{\dfrac{H_{dem}}{\eta_b} + \dfrac{E_{PM}}{\eta_{e,pp}}} \qquad (1\text{-}4)$$

$$\pi = \frac{E_{PM} + \dot{\phi}_H}{\dot{\phi}_{in,CCHP}} \qquad (1\text{-}5)$$

In the above equations it is assumed that the recoverable heat from the prime mover satisfies the heat demand (H_{dem}) of the building, and E_{PM} is the kW of electricity that is generated by the prime mover under full load or partial load operation. In addition, H and η stand for heating and efficiency, respectively. Also the subscripts of PM, in, pp, dem, b, and e mean prime mover, input, conventional power plant, demand, boiler, and electrical, respectively. Moreover, H_{dem} represents the heating load plus DHW load. In addition if there is simultaneous heating and cooling, H_{dem} represents summation of the heating load, the DHW load, and the heat required to operate the absorption chiller. The authors concluded that because the *FESR* and π consider different magnitudes and types of energy demand, they can show the real behavior of the CCHP system much better.

However, the above criteria are not enough for designing a CCHP system, because these criteria do not consider the impact of different energy management methods.

Reference [3] compares two energy management strategies of *thermal demand management* (TDM) and management of *primary energy savings* (PES). The first method is also called *following thermal demand* (FTL) in which preparing the thermal demand is the top priority for the CCHP system and must be satisfied by the designed CCHP system; this means that if the electricity production of the CCHP system does not fulfill the electrical demand of the building, the lack of electricity would be compensated for by the electricity grid. The PES method, which is the main proposal of Ref. [3], sets a constraint to ensure the energy savings of the CCHP in the working period as follows:

$$Fuel_{CCHP} \le \frac{H_{dem} \cdot \eta_{e,pp}}{\eta_b(\eta_{e,pp} - \eta_{e,CCHP})} \tag{1-6}$$

Based on the PES strategy proposed by [3] the engine operates in full load as long as the FESR is positive and energy saving occurs.

The above equation does not consider the grid loss. Since the electricity loss due to transmission and distribution (T&D losses) of electricity is significant and may reach as high as 10% in some countries, we recommend counting it. Therefore for better judgment the following equation is suggested for yearly calculations when the CCHP system follows the thermal load (FTL):

$$\sum_{yearly} F_{pp} + \sum_{yearly} F_b \ge \sum_{yearly} F_{CCHP} \tag{1-7}$$

Substituting the equivalent expressions in the above equation and bearing in mind that the CCHP considered here follows the FTL strategy results in the following:

$$\sum_{yearly} \frac{E_{PM}}{\eta_{e,pp}\eta_g} + \sum_{yearly} \frac{H_{dem}}{\eta_b} \ge \sum_{yearly} F_{CCHP} \tag{1-8}$$

$$\sum_{yearly} \frac{\eta_{e,CCHP}F_{CCHP}}{\eta_{e,pp}\eta_g} + \sum_{yearly} \frac{H_{dem}}{\eta_b} \ge \sum_{yearly} F_{CCHP} \tag{1-9}$$

and finally

$$\sum_{yearly} F_{CCHP} \le \frac{\displaystyle\sum_{yearly} H_{dem}}{\left(1 - \dfrac{\eta_{e,CCHP}}{\eta_{e,pp} \cdot \eta_g}\right)\eta_b} \tag{1-10}$$

In the above equation it is assumed that the prime mover electrical efficiency remains constant in PLO and is equal to its nominal efficiency.

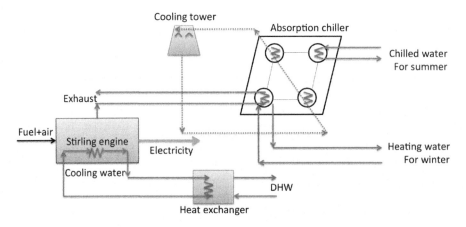

Figure 1.3 Schematic of the Stirling engine–based CCHP system proposed by [4].

Reference [3] used the maximum rectangle method (MRM) to size the CCHP components. In 2004, Ref. [4] presented some simulations for a CCHP system that was designed based on a Stirling engine. The cycle is shown in Figure 1.3.

As can be seen this cycle uses an absorption chiller for cooling purposes. The exhaust heat is used to run the chiller in cooling mode and prepare heating load for the summer mode. The base load, which is the DHW load, is prepared in a heat exchanger by the cooling water of the compression side of the Stirling engine. To investigate the applicability of this CCHP, primary energy savings (PES), primary energy rate (PER), and the economic criteria annual saving cost (AS), annual avoided cost (AC), and payback period (PB) are evaluated. The PER is defined as follows:

$$PER_{CCHP} = \frac{\sum_{yearly} F_{CCHP}}{\sum_{yearly} (H_{CCHP} + C_{CCHP} + DHW_{CCHP} + E_{PM})} \tag{1-11}$$

Hence, the PER is the ratio of annual fuel consumption of the CCHP system to the annual summation of useful energy outputs from the CCHP system including heating, cooling, DHW, and electricity. It is clear the smaller the PER the better the CCHP system.

Annual avoided cost (AC) is defined as the summation of all costs that are not paid when using CCHP system but are paid when using SCHP:

$$AC = \sum_{yearly} (i_C C_{CCHP} + i_H H_{CCHP} + i_{DHW} DHW_{CCHP} + i_E E_{PM}) \tag{1-12}$$

where i is the index price of buying energy. Reference [4] defines AS as follows:

$$AS = AC - I_{OM} - \sum_{yearly} i_{Fuel} F_{CCHP} \tag{1-13}$$

In the above equations selling electricity to or buying electricity from the grid is not considered. A CCHP that is able to sell the surplus electricity to other consumers or the grid makes a significant profit and can change the magnitude of avoided costs and annual saving costs considerably. Also it may be necessary to sometimes buy electricity from the grid, such as during a scheduled or nonscheduled shutdown or when the optimum nominal size of the prime mover is smaller than the electrical peak load. Therefore in designing a CCHP system buying and selling electricity should be considered as parameters that have a significant impact on economic evaluation criteria such as AC, AS, and PB, which are considered in Ref. [4].

In 2005, Ref. [5] analyzed a CCHP system that was designed based on a micro-gas turbine (MGT) fueled with biogas. They used different configurations of several MGTs with single-, double-effect water/LiBr and ammonia/water absorption chillers. The exhaust gas is supposed to be used directly or indirectly in the chiller. The chillers are also able to work with direct fire, therefore the authors defined a parameter called post combustion factor (PCF) as follows:

$$PCF = \frac{F_{MGT}}{F_{MGT} + F_{abc}} \tag{1-14}$$

If PCF equals zero, it means that the MGT is off and the chiller may be working with direct fire only; however when PCF equals one, the chiller may be working and only uses the exhaust gas of the MGT. In addition if PCF is between zero and one, it means that the chiller is using both the exhaust gas and direct fire.

Different research has shown varying interest in using cooling technologies and prime movers. Reference [6] in 2006 used the exhaust gas of an MGT to produce high-pressure primary motive flow of a vapor jet ejector to make suction and pressurize the secondary suction flow of refrigerant in the ejector cooling cycle. The refrigerants that were used include water, ammonia, and HFC-134a. In fact in the presented cycle, the exhaust gas runs the vapor jet ejector, which plays the compressor role in the cooling cycle. A representation of the cycle is shown in Figure 1.4.

They calculated the COP of the ejector cooling cycle as follows:

$$COP_{ej} = \omega \frac{h_2 - h_1}{(h_5 - h_4) + (h_4 - h_3)} \tag{1-15}$$

where ω is called the entrainment ratio and equals the mass flow rate ratio of secondary suction flow to the primary motive flow ($\dot{m}_2 / \dot{m}_5$). Designing the optimum value of the entrainment ratio has a significant impact on the cycle efficiency.

Reference [7] developed a code to determine the size of a small-scale CHP system to produce the electricity and heat required for heating or cooling. This code is able to perform thermo-economical and environmental analyses. It presents a general procedure to specify the type, size, and number of CHP units required for a special purpose. To determine the type and size of the prime mover, the electrical, thermal, and cooling loads of the consumer are calculated for different time periods according to the energy

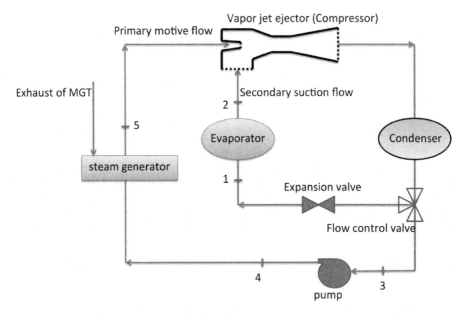

Figure 1.4 Ejector cooling cycle based on the exhaust gas of an MGT proposed by [6].

bills and calculations, and the average power to heat ratio ($\overline{PHR}$) of the consumer is calculated as follows:

$$\overline{PHR} = \frac{\sum_{i=1}^{N}\left(\frac{kW_e}{kW_{th}}\right)_i}{N} \tag{1-16}$$

where N is the number of time periods that the consumer loads are calculated or measured. Attention must be paid to the calculation of kW_{th}. It is the total heat required for heating or cooling if a consumer uses an absorption chiller.

In this research ([7]) the main criteria for sizing and selecting the prime mover are the thermal load curve, $\overline{PHR}$, and energy management strategies. This code is able to evaluate the parameters for two strategies: power match or thermal match. In the power match strategy the prime mover is designed to satisfy the maximum electrical demand of the consumer, while in the thermal match strategy, which is also called following thermal load (FTL), the prime mover is designed to satisfy the maximum thermal demand (including the heat for DHW and cooling) by using the recovered heat from the prime mover.

In the economic evaluations some important criteria such as the net present value (NPV), internal rate of return (IRR), payback period (PB) and benefit to cost ratio (BCR) are calculated. Another important economic criterion is the cost per kWh_e produced by the electricity production unit. In CHP or CCHP units, however, some of the operation and maintenance (O&M) and fuel costs are used to produce heating or cooling (ex'), therefore in calculating the cost per kWh_e the ex' must be subtracted from the total annual cost of the system (ex). Reference [7] called this parameter the *levelized electricity production cost* (LEPC). Since some parameters fluctuate the average value of LEPC is calculated as follows:

$$\overline{LEPC} = \frac{1}{L}\sum_{y=1}^{L}\frac{1}{(1+r)^y}\frac{(ex-ex')_y}{(total\ kWh_e\ production)_y} \tag{1-17}$$

where r, y, and L are the interest rate, year of operation, and lifetime of the project, respectively.

Reference [7] also calculated the FESR as follows:

$$FESR = 1 - \frac{\dfrac{\overline{\eta}_{e,pp}}{\eta_{e,CCHP}}}{1+\left(\dfrac{\overline{\eta}_{e,pp}}{PHR.\eta_b}\right)} \tag{1-18}$$

where the average electrical efficiency of conventional power plants ($\overline{\eta}_{e,pp}$) is

$$\overline{\eta}_{e,pp} = \frac{1 - PerCont_{no-fos}}{\sum_{l=1}^{M}\left(\dfrac{PerCont_{fos}}{\eta_{e,pp}}\right)_l} \tag{1-19}$$

where M and $\eta_{e,pp}$ are the total number of different types of power plants and the electrical efficiency of the particular power station, respectively. In addition $PerCont_{fos}$ and $PerCont_{no-fos}$ are the percentage of contribution of fossil fuel and nonfossil fuel power stations, respectively. The percentage of contribution is calculated using the MW of power generation by the particular type of power plant in relation to the total MW generated by all of the power plants.

Two review articles [8, 9] published in 2006 studied small and micro-CCHP systems, equipment used, products, and field test results. In the technical data of this work, economic information and environmental pollution production of different types of prime movers and cooling and heating systems are presented. The prime movers that are studied in these papers include conventional steam and gas turbines, MGTs, Stirling engines, fuel cells, reciprocating internal combustion engines, and SCHPs.

The cooling systems that are studied include single-, double-, triple-effect absorption chillers, adsorption chillers, electric vapor compression chillers, and desiccant dehumidifiers.

Some of the technical, economic, and environmental data of the prime movers that are presented in [8, 9] include electrical efficiency; overall efficiency; performance in partial load operation; emission indices of CO_2, CO, and NO_x; reliability; availability;

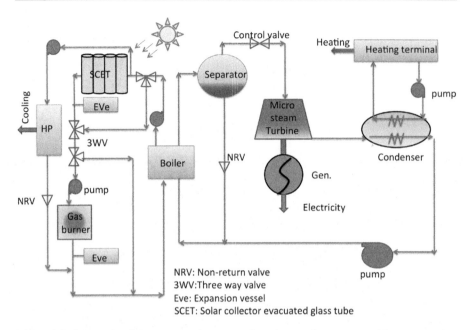

Figure 1.5 Schematic of the cogeneration system based on a micro-steam turbine proposed by [10].

capacities; fuel used; PHR; investment cost; operation and maintenance cost; lifetime; noise; and recoverable heat source temperature. A system is 100% reliable if unscheduled shutdown does not occur. Availability of a system is measured according to the percent of time the system operates with respect to the time the system needs to operate properly.

Reference [10] presented an experimental work based on a Rankine cycle to produce heat and electricity. This cycle can also be used for cooling production by using adsorption chillers that need a low-quality heat source. A schematic of this cycle is redrawn and presented in Figure 1.5.

As Figure 1.5 shows, the CHP proposed by [10] can be upgraded to a CCHP system by using the heat pump or using the recovered heat from the condenser in an adsorption chiller. Figure 1.6 represents the experimental setup of the installed CHP system from Ref. [10].

The setup shows the main components of the CHP cycle described above. They include a 25 kW Thermomax evacuated glass tube solar collector that uses an Alfal Laval Ltd vacuum blazed heat exchanger, a 1.5 kWe radial flow micro-steam turbine/generator (TTL Dynamics) with rotational speed of 60,000 rpm and frequency of 1 kHz, a 23 kW boiler working at 90°C with volume of 11 liters, a vacuumed blazed condenser with volume of 3.5 liters and heat discharging capacity of 21 kW at 35°C, a separator with volume of 50 liters, an explosion-proof electric pump (HFE-301) from GE Motors and Industrial Systems, and a double diaphragm pump (n-pentane) driven by compressed air from Aroplus Ltd.

Figure 1.6 Experimental setup of the installed CHP system [10].

The working fluid in the left side of the CHP cycle in Figure 1.5 is water but the working fluid in the right side of the cycle is n-pentane or HFE-301. Finally, n-pentane, due to having higher flammability, higher boiler operating temperature, smaller turbine isentropic efficiency, and lower steam quality is rejected and HFE-301 is chosen as the main working fluid of the Rankine cycle.

Economic evaluation of CCHP systems is of great importance. Since we are fairly sure about the emission and fuel consumption reductions of CCHP systems with respect to SCHP systems, an economically profitable CCHP is likely to become increasingly popular. Reference [11] presented a comprehensive study on the economic criteria that can be used for economic evaluations of CHP and CCHP systems. A procedure proposed by [11] for economic evaluations and decision-making about different projects is further explained and extended in the following:

1. Consider a set of nominated projects for investment.
2. Define the economic criteria that you are looking to calculate for the analyses.
3. Define the time period of the economic analyses; the period is usually limited by the working lifetime of the project.

4. Estimate the positive and negative cash flows for each project; these cash flows consist of initial investment cost, I_{OM}, and all economic expenses and income of the project such as selling electricity, not buying electricity, buying electricity, buying fuel, and so on.
5. Specify the interest rate of your economic calculations (r); this is necessary for many economic criteria such as NPV, IRR, and minimum attractive rate of return.
6. Compare the economic criteria defined in step 2 for nominated projects in order to sort the projects from very good to very bad.
7. Accept or reject the nominated projects according to the sorting in the previous step.

Table 1.1 summarizes the criteria that are discussed in Ref. [11]. In this table, I, SV, PV, cf_y and $\overline{cf}$ stand for initial investment cost; salvage value of the project after its lifetime (L); present value; yearly cash flow, which subtracts the annual expanses (ex) from the annual earnings (er) of the project; and average cash flow during the project lifetime. In addition the following definitions are used in Table 1.1:

$$FV(income) = income.(1+r)^{L} \tag{1-20}$$

$$PV(cf_x) = \frac{cf_x}{(1+r)^x} \tag{1-21}$$

$$PV(er) = \sum_{y=1}^{L} \frac{er_y}{(1+r)^y} \tag{1-22}$$

$$PV(ex) = \sum_{y=1}^{L} \frac{ex_y}{(1+r)^y} \tag{1-23}$$

$$PV(SV) = \frac{SV}{(1+r)^{L}} \tag{1-24}$$

Table 1.1 The Useful Criteria for Economic Evaluation of CCHP Systems [11]

Classification	Criteria	Formula	Use and Constraints
The NPV methods	Net present value (NPV)	$NPV = -I + \dfrac{SV}{(1+r)^{L}} + \sum_{y=1}^{L} \dfrac{cf_y}{(1+r)^y}$	Gives the present value of all cash flows of the project during its lifetime. A project is acceptable when NPV > 0.
	Future value (FV)	$FV = NPV.(1+r)^{L}$	Presents the future value of the NPV after the lifetime of the project.
	Annual value (AV)	$AV = \dfrac{NPV}{\dfrac{1-(1+r)^{-L}}{r}}$	AV converts NPV to a series of usually equal cf.

Table 1.1 **The Useful Criteria for Economic Evaluation of CCHP Systems [11]** *(Cont.)*

Classification	Criteria	Formula	Use and Constraints
The rate of return methods	Internal rate of return (IRR)	$-I + \dfrac{SV}{(1+IRR)^L} + \sum\limits_{y=1}^{L} \dfrac{cf_y}{(1+IRR)^y} = 0$	IRR is the interest rate that sets NPV equal to zero. A project is acceptable when IRR $> r$.
	External rate of return (ERR)	$ERR = \sqrt[L]{\dfrac{FV(income)}{I}} - 1$	ERR calculates the interest rate for which the FV of the initial investment equals the FV of all other cash flows invested at the minimum attractive rate of return.
	Growth rate of return (GRR)	$GRR = \sqrt[x]{\dfrac{PV\,(cf_x)}{I}} - 1$	This equation calculates the rate of return of project at a particular year (x) of the project's lifetime. NPV must be positive.
The ratio methods	Premium value percentage (PVP)	$PVP = \dfrac{NPV}{I}$	Determines the net profit from every unit of money (1 USD for example) invested.
	Return on original investment	$ROI = \dfrac{\overline{cf}}{I}$	Determines the average annual return from the initial investment. NPV must be positive.
	Return on average investment	$RAI = \dfrac{\overline{cf}}{\overline{I}}$	Determines the average annual return from the average initial investment.
	Benefit to cost ratio (conventional)	$BCR == \dfrac{PV(er) - PV(ex)}{I - PV(SV)}$	A project is acceptable when $BCR > 1$.
	Benefit to cost ratio (Lorie-Savage)	$BCR' == \dfrac{PV(er) - PV(ex) - I}{I - PV(SV)}$	A project is acceptable when $BCR' > 0$.
	Profit to investment ratio (PIR)	$PIR = (\sum\limits_{y=1}^{L} cf_y - I)/I$	Determines the profitability of the project with respect to the initial investment. PIR must be positive.
Payback methods	Payback period (PB)	$PB = \dfrac{I}{\overline{cf}}$	Calculates the years it will take to recover the original investment. NPV must be positive.

MGTs use a recuperator to recover the heat of high-temperature (500–540°C) exhaust gases from the micro-turbine to preheat the compressed air before it enters the micro-combustion chamber. Although MGTs use the recuperator, the exhaust gas temperature leaving the recuperator is typically in the range of 250–300°C; this can be recovered again for cooling by utilizing heat-driven chillers, heating, or bottoming micro-organic Rankine cycles that can work with low-quality heat sources.

Reference [12] investigated a bottoming micro-organic Rankine cycle (micro-ORC) that used the exhaust of an MGT to increase the energy utilization of the cycle by increasing electricity generation. They proposed a procedure to select the best working fluid from environmental and technical points of view. For this purpose 16 fluids were considered and compared according to their impact on the environmental and technical criteria. A schematic of the MGT-ORC is drawn and presented in Figure 1.7.

Reference [13] presented a method for sizing and choosing the number of identical Capstone C30 MGTs required for meeting the cooling, heating, and electricity demands of a building in three cities in Iran. A representation of the cycle is redrawn and integrated here in Figure 1.8. The cycle consists of MGT units, an absorption chiller, a heat pump, an auxiliary heater, and a heat exchanger for heat recovery. They used the energy balance method to satisfy the cooling and heating loads of a building by recovering the exhaust energy from the MGTs when determining the number of necessary MGT units. The final number of MGT units is the maximum

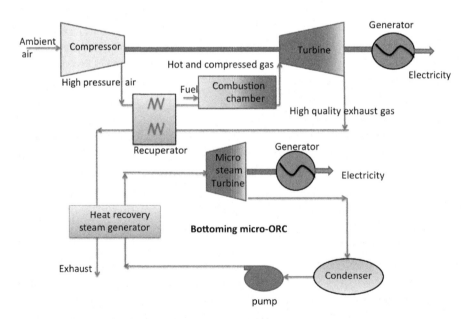

Figure 1.7 Schematic of the combined MGT-ORC proposed by [12].

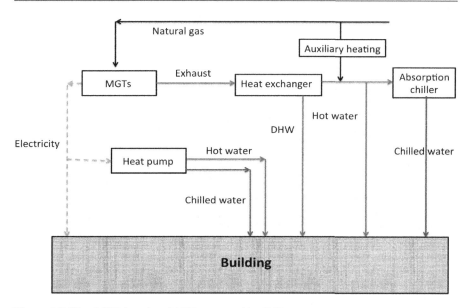

Figure 1.8 The CCHP based on MGTs proposed by [13].

number proposed by the balance energy equations. The following equations present the method:

$$(NOU_{win}E_{PM} - E_{dem})COP_{hp-heating}$$
$$+NOU_{win}Q_{rec}\varepsilon_{HE} - D_{dem} = H_{dem} \quad \text{, for winter}$$ (1-25)

$$(NOU_{sum}E_{PM} - E_{dem})COP_{hp-cooling}$$
$$+(NOU_{sum}Q_{rec}\varepsilon_{HE} - D_{dem})COP_{abc} = C_{dem} \quad \text{, for summer}$$ (1-26)

$$\rightarrow NOU = \max(NOU_{win}, NOU_{sum})$$ (1-27)

If the authors of Ref. [13] plan to use an auxiliary heater as they have shown in the schematic of the cooling cycle presented, the energy balance equations should be rewritten as follows:

$$(NOU_{win}E_{PM} - E_{dem})COP_{hp-heating}$$
$$+NOU_{win}Q_{rec}\varepsilon_{HE} + Q_{aux,win} - D_{dem} = H_{dem}$$ (1-28)

$$(NOU_{sum}E_{PM} - E_{dem})COP_{hp-cooling}$$
$$+(NOU_{sum}Q_{rec}\varepsilon_{HE} + Q_{aux,sum} - D_{dem})COP_{abc} = C_{dem}$$ (1-29)

The energy consumed by the auxiliary heater (Q_{aux}) may be different in summer and winter. In addition the number of units (NOU) of the MGTs may be reduced if an

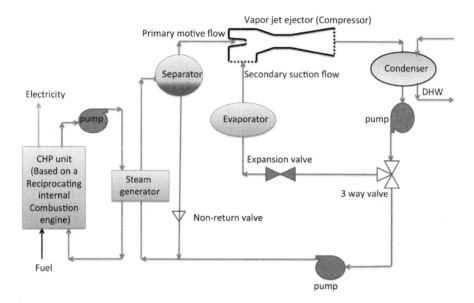

Figure 1.9 The CCHP based on a Dachs CHP unit and ejector refrigeration cycle proposed by [14].

optimization procedure could be applied to optimize the number of MGTs and auxiliary heater capacity.

Reference [14] presented a mathematical and experimental investigation of a CCHP system that was based on a reciprocating internal combustion engine. The engine is used as a CHP unit (manufactured at Dachs with 5.5 kW_e and 12.5 kW_{th} outputs) that provides the heat required for steam generation in a vapor jet ejector refrigeration cycle. The recovered heat as hot water at a temperature of about 83°C is supplied to the steam generator. Heat is recovered from the generator, lube oil, water jacketing, and exhaust gases. A separator is installed before the ejector to separate liquid from the steam. The schematic of the cycle is redrawn in Figure 1.9.

The refrigerants that were nominated for the ejector cooling cycle were water steam, HFE 7100, and HFE 236fa. Steam was rejected due to its need for high superheating to avoid droplet formation in the entrance of the ejector, and HFE 236fa was also rejected due to the need for higher working pressure in the ejector. HFE 7100 was finally chosen as the working fluid in the ejector refrigeration cycle.

The main evaluation functions defined in [14] include η_{CCHP} and CO_2 emission ratio (R):

$$\eta_{CCHP} = \frac{E_{PM} + Q_{CCHP} + C_{CCHP}}{F} = \frac{E_{PM} + xQ_{CHP} + (1-x)Q_{CHP}COP_{ej}}{F} \qquad (1\text{-}30)$$

where $x = Q_{CCHP} / Q_{CHP}$ is the ratio of the net output heat from the CCHP system to the net output heat from the CHP unit. According to the cycle Q_{CCHP} equals the heat used for preparing the DHW.

The CO_2 emission ratio is defined in the following; when R is greater than 1, the CCHP reduces emissions:

$$R = \frac{Em_{SCHP}^{CO_2}}{Em_{CCHP}^{CO_2}} \tag{1-31}$$

Reference [15] investigated the economic feasibility of combined heat and power with different prime movers such as MGT, gas turbines, and diesel engines in a liberalized energy market. The main parameter used for economic evaluation in this study is the NPV, which is defined as follows:

$$NPV = -I + \sum_{y=1}^{L} \frac{cf_y'}{(1+r)^y} \tag{1-32}$$

where cf' is the modified yearly cash flow when taxation impact is also considered:

$$cf_y' = cf_y - \left(cf_y - \frac{|I|}{L} \right) \cdot tax \tag{1-33}$$

where the *tax* is the taxation rate for the net annual profit.

A micro-CCHP based on a reciprocating internal combustion engine working with liquefied petroleum gas (LPG) and natural gas was proposed by Ref. [16]. The cooling system was an adsorption chiller and the heating system used hot water in the floor heating system to provide comfortable conditions in the building. The exhaust and water jacketing heat are recovered through an exhaust heat exchanger (EHE) and a plate heat exchanger (PHE). The authors' analyses include energy, exergy, and economic (NPV and PB) evaluations.

A CCHP system proposed in Ref. [17] (2008) is based on a gas turbine to produce electricity and recoverable heat. The recovered heat is either used in an absorption chiller to produce chilled water or for heating purposes. An auxiliary gas boiler is used to compensate for the lack of heat for heating. In addition, an electric chiller is used to compensate for the lack of cooling production in the absorption chiller. The system can receive electricity from the grid in case of lack of electricity production by the gas turbine as well. The purpose of the study was to minimize an annual total cost function that includes the fixed costs and variable costs as follows:

$$ex = fixed\ cost + variable\ costs \tag{1-34}$$

The fixed costs may include annual fixed charges for regular checks and maintenance work. The variable costs include fuel costs, electricity costs, and variable maintenance costs.

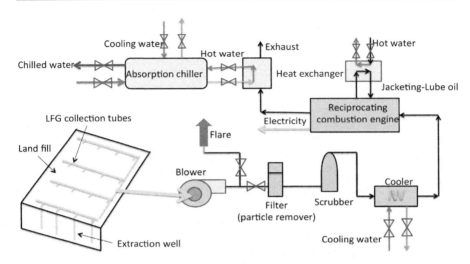

Figure 1.10 The schematic of the LFG-CCHP system [18].

Reference [18] proposed a landfill gas (LFG) CCHP system. The LFG-CCHP system consists of three main subsections: the LFG collection system, the LFG processing system, and the CCHP system itself. The schematic of the LFG-CCHP system is redrawn and presented in Figure 1.10.

As Figure 1.10 shows the LFG is collected through pipes that are connected to a blower. The blower produces a small suction pressure and LFG flows through the collection pipes. Then a filter removes particulate pollution, and a scrubber cleans the chlorine, sulfur, and ammonia content from the LFG. The water vapor is condensed, separated, and drained in the cooler. The extra LFG will be flared in the flare line.

In the CCHP system exhaust heat is used for cooling generation in an absorption chiller while the other heat sources such as water jacketing and the lube oil cooler are used for hot water production.

The composition of the LFG includes 50% methane, 45% carbon dioxide, 5% nitrogen, and less than 1% oxygen. Other components include hydrogen sulfide (H_2S), halides, and nonmethane organic compounds [18]. NPV is calculated in the economic evaluations. PES is calculated for the energy consumption evaluation and greenhouse gas (GHG) reduction is determined for the environmental assessment.

Up to this point whenever a CCHP cycle is proposed there has been no mention about how the component types are chosen. For example, if a CCHP cycle is created based on a MGT, what would happen if a reciprocating internal combustion engine is used instead of the MGT?

Many factors, such as the economic criteria, thermal efficiency, electrical efficiency, environmental impact, maintenance jobs and cost, and so on, will change by changing only the type or even size of the prime mover, cooling system, or heating system.

References [19] and [20] used two multicriteria decision-making methods (MCDM) for choosing the best set of components to build a CCHP system among various alternatives. They used fuzzy logic and the grey incidence approach for the decision-making process. In these methods all the alternatives are compared from different points of view such as economic, environmental, technological, social, etc. The criteria that the alternatives are compared with are numerous and diverse. MCDM methods integrate all of these numerous and different criteria in one single number to make the decision-making more straightforward.

Economic, thermodynamic, and environmental evaluations of a CCHP based on different energy management strategies are undertaken by Ref. [21]. They considered three strategies: FTL, FEL, and a hybrid electrical thermal strategy (HETS). In HETS, according to the thermal or electrical load the CCHP switches between the FTL and FEL. They concluded that the last strategy creates a good balance between the evaluation criteria. Since optimizing one criterion may increase or decrease the other criteria, they defined a function to integrate the three criteria of annual primary energy consumption (PEC), annual operation costs (*ex*) and annual carbon dioxide emission production. This function is called the *performance factor indicator (PFI)* and is written as follows:

$$PFI = \frac{PEC_{CCHP}}{PEC_{SCHP}} + \frac{ex_{CCHP}}{ex_{SCHP}} + \frac{Em_{CCHP}^{CO_2}}{Em_{SCHP}^{CO_2}} \tag{1-35}$$

In this function no weight is defined for the criteria and the PFI is not normalized. According to the definition of PFI, the smaller the PFI is the better the CCHP system will be. The CCHP cycle that is analyzed in Ref. [21] is regenerated in Figure 1.11.

In the FEL strategy, the prime mover output electricity equals the consumer electrical demand. This means that the prime mover should operate at partial load. Working at partial load decreases the electrical efficiency of the prime mover, especially if it is a gas turbine type. Moreover operation of prime movers below the minimum design load is impossible because it results in bearing damage due to metal to metal contact, vibration, and surge (in the compressor of gas turbines), and finally forced shutdown

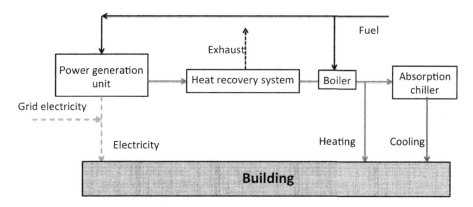

Figure 1.11 The CCHP proposed by [21].

of the prime mover. Therefore when designing a CCHP system, defining a load-dependent electrical efficiency and a minimum working load for the prime mover helps to achieve more realistic results. Consequently in working loads below the minimum working load of the prime mover, electricity should be provided form the grid. This also changes the economic evaluations of the CCHP cycle. Another parameter that should be clearly defined in the analyses is the operation mode of the prime mover. The time when the prime mover is supposed to be working at full load or partial load or not working at all should be defined. If a prime mover works at full load during the whole year, the magnitude of the evaluation criteria is totally different than that of a prime mover that works at partial load.

Reference [22] proposed a fitness function (*ff*) that should be optimized (maximized) by using the genetic algorithm to find the optimum configuration of component sizes. The fitness function is defined as follows:

$$ff = \omega_1 \cdot FESR + \omega_2 \cdot ATCSR + \omega_3 \cdot CO2RR \qquad (1\text{-}36)$$

In the above equation, fuel energy saving ratio (FESR), annual total cost saving ratio (ATCSR), and carbon dioxide emission reduction ratio (CO2RR) are considered and equal weights are given to the criteria. The cycle analyzed by Ref. [22] is presented in Figure 1.12.

Reference [23] presented a CCHP system approximately similar to that presented in [21] except that electricity sold back is also assumed in this study. In this research they defined three different functions of annual operational cost (*ex*), primary energy

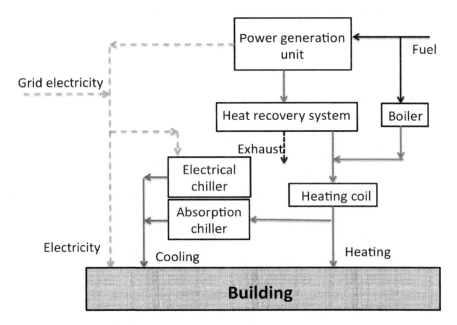

Figure 1.12 The CCHP cycle proposed by [22].

consumption (PEC), and CO_2 emission production (Em_{CO_2}). They minimized these functions and concluded that optimizing these three functions at the same time is not possible for all of the climates they considered for the study.

Reference [24] also presented a CCHP system based on an MGT for a supermarket and examined FTL and FEL strategies. In the evaluations they calculated PB with respect to two parameters: the percent of the recoverable heat of the MGT that is used in the absorption chiller, and the electricity to gas price ratio. They also calculated overall efficiency (η_o), CO_2 emission, and FESR with respect to the percent of the recoverable heat of the MGT that is used in the absorption chiller to produce cooling. In the calculation of the criteria the availability of the MGT is also considered.

A super-structure was proposed by [25] for economic evaluation of a CCHP system. In this super-structure they considered two alternatives, a reciprocating internal combustion engine and an MGT, as the prime movers. They also considered an electrical chiller, single- and double-effect absorption chillers, and a cooling tower (as the heat remover for the cooling water of absorption chillers) as the possible options for the cooling system. The heating systems that are considered include a steam boiler, hot water boiler, vapor–hot water heat exchanger, hot water–cooling water heat exchanger, heat recovery in the MGT, and an internal combustion engine.

The economic criterion that is considered to find the optimum configuration is the total annual cost, which includes the fixed and variable costs. This function should be minimized. Similar approaches can also be found in [26-27]. Reference [26] minimized the variable annual cost function and Ref. [27] maximized annual cash flow (cf) to find the optimum size and unit numbers of MGTs for determined thermal and electrical load curves.

Since building demands are variable, the prime mover of a CCHP system may be working due to the need for electricity while no heating or cooling is required. In this case the recoverable heat of the prime mover can be stored in the thermal storage systems; otherwise it will be lost. In another scenario the building may need cooling or heating while no electrical energy is needed. In this case the prime mover should operate and sell the electricity to the gird or use backup equipment for cooling or heating. Therefore some tasks should be defined for each piece of equipment according to the building demands and the criteria on which the CCHP is based. This means the CCHP should choose a task in order to optimize the design criteria. This topic is discussed in Ref. [28]; they used a "task configuration system" for a CCHP system including an MGT, absorption chiller, electrical chiller, and thermal storage tank for preparing hot water for heating demands.

A CCHP system must have two other important characteristics: first it should be feasible, and second it should be flexible. A CCHP system is feasible when it is able to support consumer demands at any time. The flexibility is concerned with the ability of the CCHP to work with changeable loads while keeping its overall efficiency as high as possible. The more changeable the consumer demand is, the more flexible the CCHP system should be. Feasibility should be satisfied prior to flexibility. Among the prime movers for example, reciprocating internal combustion engines are more flexible as compared to gas turbines and MGTs.

References [29] and [30] look at the feasibility and flexibility of the CCHP systems. They calculated the feasibility and flexibility of some CCHP systems that were based on the simplified CCHP that is shown in Figure 1.13. They also evaluated the

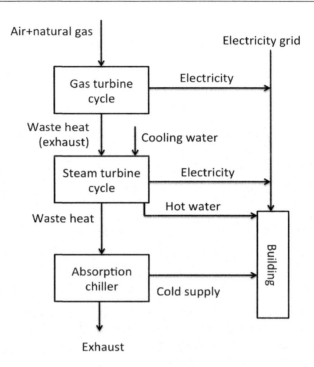

Figure 1.13 The general CCHP studied by [29].

feasibility and flexibility of the CCHP system when integrated with options such as a thermal storage system, additional water heater, or electric heater.

In other research, different configurations of MGTs and single-, double- effect water/LiBr, ammonia/water absorption chillers were studied when the system is fueled with biogas and natural gas [31]. The biogas composition used in this study included 64%, 31.4%, 3%, and 0.2% of CH_4, CO_2, H_2O, and H_2S respectively. It also includes 100 ppm of siloxanes. They used thermoeconomic and environmental analyses for different configurations and finally chose the best one. The cooling produced by the absorption chillers is used for space cooling and/or a biogas pretreatment process.

Reference [32] coupled an organic Rankine cycle with an ejector refrigeration cycle for use as a CCHP system. The working fluid of the ORC is the R123, which is nontoxic, nonflammable, and noncorrosive. The solar energy is captured by a compound parabolic collector (CPC); it is stored in thermal storage and later reheated by an auxiliary heater for steam generation in a boiler (Figure 1.14). R123 enters the boiler and the generated steam runs the steam turbine and the ejector cooling cycle. An extraction is taken from the turbine that can be used for heating or cooling purposes.

They optimized the cycle with a genetic algorithm from the exergy point of view while the CPC orientation and position are optimized.

References [33] and [34] propose procedures to find the best refrigerant for the ORC. References [35] to [39] also present thermoeconomic and environmental evaluations for different CCHP cycles and applications. Reference [35] proposed a fitness

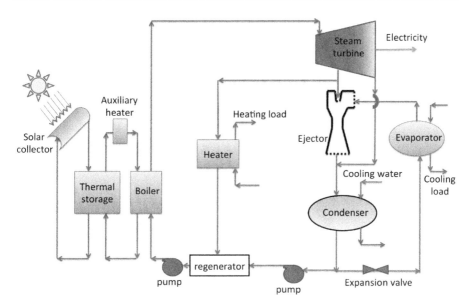

Figure 1.14 Ejector cooling ORC-CCHP working with solar energy.

function (*ff*) similar to that presented by [22] with equal weights for FESR, ATCSR, and CO2RR. The *ff* is optimized by genetic and particle swarm optimization algorithms and the results are compared. The parameters that were considered for optimization in order to maximize the *ff* include the prime mover size, thermal storage capacity, on-off coefficient of the prime mover, and ratio of electric cooling to cooling demand. The energy flow of the cycle is depicted in Figure 1.15.

The CCHP system proposed by [36] was a turbine-driven system; they considered three strategies in the evaluations: FEL, FTL, and FETS. In the FETS a load ratio (LR) is defined in order to specify the working mode of the CCHP system. The LR is defined as follows:

$$LR = \frac{Monthly\ electric\ load}{Monthly\ thermal\ load} \tag{1-37}$$

According to this strategy if LR > 1 the system should use FEL in that month; otherwise it should operate based on the FTL.

Carbon credit (*cc*) is a law that can make CCHP systems more popular, because they reduce CO_2 production. Based on the carbon credit law, if the carbon production by the CCHP is smaller than a reference set value, a savings due to carbon production reduction is achieved by the CCHP system. It is calculated as following [36]:

$$savings_{CCHP-carbon} = (Em_{ref}^{CO_2} - Em_{CCHP}^{CO_2})cc \tag{1-38}$$

where the *cc* is measured based on $/(metric ton of CO_2 equivalent).

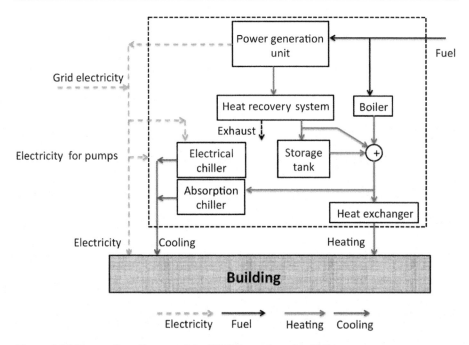

Figure 1.15 Energy flow diagram of the CCHP investigated in [35].

A CCHP proposed by [37] is based on an ORC-CHP that uses electricity to produce cooling in an electric vapor compression chiller as presented in Figure 1.16. They also investigated several refrigerants for use in the ORC. The main criteria for the evaluation was annual PEC, ex, and Em_{CO2}.

Reference [38] proposed a CCHP where the prime mover used biogas as its fuel. The recovered heat from the prime mover and an auxiliary boiler that runs on natural gas provides required heat for heating or cooling via an absorption chiller. The system can sell or purchase electricity to or from the grid. A total cost function including the summation of the investment costs, variable operation and maintenance, bought electricity, biogas, natural gas, and carbon tax minus the sold electricity is considered for minimization.

Reference [40] also presented an experimental work about an ORC-CHP cycle. The ORC is used to recover the exhaust heat from a 21 kW dual fuel combustion engine and convert it to electricity. The authors evaluated the impact of the ORC-CHP cycle on the fuel energy conversion and emission production. They also analyzed the impact of pinch point temperature difference on the exergy efficiency and exergy destruction.

A MGT-based CCHP system proposed by [41] is basically similar to that presented in [6]. In this CCHP, water is converted to saturated steam in a heat recovery steam generator (HRSG); it is used for cooling through a vapor ejector refrigerator, or for heating via a heat exchanger. The authors investigated the impact of pinch point temperature difference on energy utilization factor, FESR, and thermal efficiency.

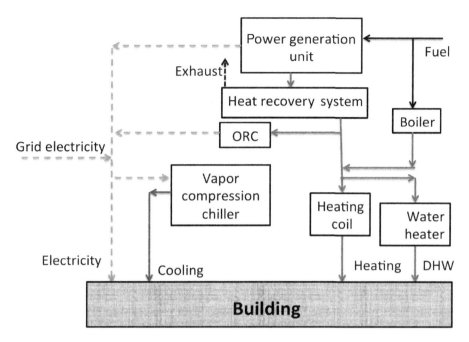

Figure 1.16 The CCHP based on an ORC-CCHP and vapor compression chiller proposed by [37].

They also considered the impact of PHR on some evaluation criteria. Furthermore, they calculated the exergy loss and exergy efficiency of the CCHP components.

Reference [42] used a particle swarm optimization algorithm (PSOA) for a CCHP system that is approximately similar to that presented in [35]. However, they considered a total cost function and minimized it with the PSOA.

The impact of energy demand uncertainty on the feasibility and design of the CCHP system is investigated in [43]. The Monte Carlo method (MCM) is used to simulate the uncertainty and is coupled with mixed-integer nonlinear programming (MINLP). The authors concluded that the impact of uncertainty on the annual saving cost rate and annual fuel saving rate is negligible. Their results also show medium sensitivity of the main equipment capacity to the uncertainty while high sensitivity is recognized for the secondary equipment.

Reference [44] considered N residential neighbors, each one having a CHP system comprised of a gas engine, storage tank, and gas-fired water heater. They compared three strategies for the N house from energy points of view (PEC). In the first strategy they considered, electricity could be interchanged between houses of the residential complex, and the system could buy electricity from the grid or any other local electricity producer to fulfill the electricity demand. In the second scenario houses are not permitted to interchange electricity between themselves; the houses used the electricity produced from the gas engine and if needed they purchased electricity from the grid. In the third scenario, all electricity provided to the houses was from the grid.

In addition, in none of the three scenarios could the surplus electricity produced by the gas engines be sold to the grid.

The maximum rectangle method (MRM) that was used for sizing of the prime mover in Reference [3] was extended to three MRM methods for horizontal, vertical, and high-level design [45]. In the horizontal design method, the engine size is the size that the classical MRM recommends, but the full-time operation is extended to the limit when the PES reaches a preassumed minimum value. In the vertical design, the full-time operation of the engine is fixed and is equal to that recommended by the classical MRM, while the engine size is enlarged until a minimum PES is achieved. Note that both the vertical and horizontal design methods are based on PES alone. In the high-level design method, both the full-time operation and engine size are enlarged until the minimum PES is achieved. There will be many cases in which PES reaches the minimum value. To choose the best among all of the cases with the minimum PES, the net present value of all of the cases with the minimum PES is calculated and the case with the highest NPV is chosen as the best size and full-time operation. In other words, in the high-level design method, the MRM is equipped with energy and economic evaluation criteria.

A thermoeconomic analysis using exergy cost and the structural coefficient of internal links was done by [46] for a multiengine, multi-heat-pump CCHP system. The hypothetical CCHP system includes two power generation units (a gas turbine and internal combustion engine, both of which are diesel-fired), two cooling systems including electricity and heat-driven chillers, a heat recovery system, a thermal storage system, and a heat pump.

In experimental research presented by [47], the recovered heat from the engine cooling and exhaust gases is used for regeneration of the silica gel desiccant wheel of an HVAC system. The micro-CHP is a gas -fired reciprocating engine and is supposed to do the tasks of the existing traditional gas boiler and the electricity grid. The authors calculated the CO2RR and PES when using the micro-CHP unit with the HVAC system, and achieved 38.6% of CO2RR and 21.2% of PES. A photograph of the air-handling unit is presented in Figure 1.17.

Figure 1.17 The air-handling unit and its desiccant wheel [47]

Another experiment [48] used a Capstone MGT with a capacity of 28 kW and a double-effect absorption chiller. The MGT is fired with propane. The high-pressure generator of the chiller is equipped with a burner working with liquefied petroleum gas (LPG); meanwhile the exhaust gas of the MGT is directly entered into the high-pressure generator. The exhaust of the high-pressure generator is reused in the low-pressure generator. This configuration lets the chiller operate even if the MGT is off, because the burner can be used independently to run the chiller. Figures 1.18 to 1.20 show the MGT, double-effect absorption chiller, and test facility, respectively.

Figure 1.18 The C30 Capstone MGT [48].

Figure 1.19 The double-effect absorption chiller [48].

Figure 1.20 The test facility test presented in [48].

Thermally driven cooling systems were investigated by [49-53]. Reference [49] analyzed the performance of a silica gel–water adsorption chiller that uses a solar water heater as the driver experimentally. A numerical simulation is also presented in [50] to consider the impact of using some models on the performance of the adsorption chillers. The results are also compared with other models. A comprehensive study about thermally activated cooling systems is presented by [51]; it covers the principals of operation, products, prototypes, problems, guidelines, and typical applications in CCHP systems. The study includes comprehensive information about LiBr/water, water/ammonia absorption chillers, adsorption chillers, and solid/liquid desiccant cooling technologies. An experimental work presented in [52] used the setup presented in [47]; in this research they investigated the performance of the HVAC system, especially the desiccant wheel and air-handling unit. Reference [53] presented a numerical simulation of a tubular solar adsorption refrigerating system using an activated carbon-methanol pair.

In addition to the building demands, climate and altitude above sea level have a great impact on the design of CCHP systems. The reason is that most equipment is designed based on the standard conditions at sea level. Using a prime mover high above sea level changes the characteristic curves of the prime mover. This is also the case for other equipment such as the boiler and absorption and adsorption chillers. The impact of height on the efficiency, power, fuel consumption, and other evaluation criteria of general CCHP equipment is studied in Ref. [54]. For example, the impact of altitude (z) on the power output of a prime mover is corrected as follows:

$$E_{PM}(z) = \xi(z)E_{nom} \qquad (1\text{-}39)$$

where $\xi(z)$ is the *altitude air density ratio*, defined as follows:

$$\xi(z) = \frac{\rho(z)}{\rho(z=0)} = [1 - 2.25577 \times 10^{-5}(z)]^{5.2559} \tag{1-40}$$

where z is measured in meters. In the above equation, the humidity is assumed to remain constant at altitude. Other equations for other equipment such as boilers, heat exchangers, pumps, absorption chillers, fans, etc. can be found in the reference. CCHP systems working with different prime movers such as steam turbines, MGTs and reciprocating internal combustion engines are also studied in [55-61]. References [58] and [59] investigated using hydrogen and biomass as fuel and compared the environmental benefits of these fuels with respect to fossil fuels. A photovoltaic (PV)-CCHP system was proposed by [62] to meet the load demands of a consumer. Figure 1.21 presents the block diagram of the proposed system. The purpose of combining the PV system with a CCHP system is to decrease the waste heat of the CHP unit, especially when the prime mover is operating at full load to provide the electricity peak.

Parallel use of an electric chiller and absorption chiller in a CCHP cycle is investigated in [63] and [64]. They investigated and optimized the impact of variation of proportion of the electrical cooling to the total produced cooling on the evaluation criteria of the cycle. They also used this strategy to determine the prime mover size. A supercritical closed gas turbine cycle working with CO_2 is proposed by [65]. This CCHP system uses solar energy and an auxiliary heater to provide the required thermal energy, and an ejector cooling cycle is used to provide the required cooling load. The schematic of the cycle is presented in Figure 1.22.

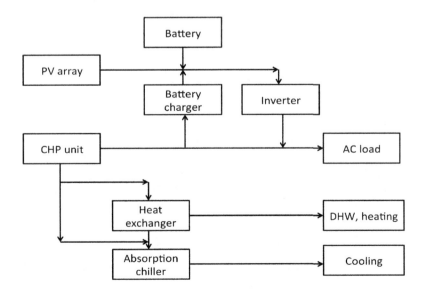

Figure 1.21 Schematic diagram of a PV-CCHP system proposed by [62].

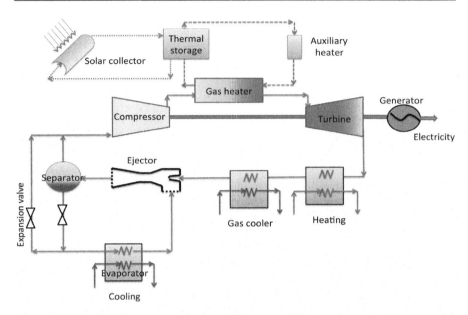

Figure 1.22 The CCHP cycle introduced in [65]

Selecting the best type of prime mover was the concern of [66] and [67]. They used multicriteria decision-making methods such as the fuzzy method. In Ref. [68] an ORC is integrated with an MGT and uses its exhaust energy to produce extra power. An ejector cooling system combined with an ORC is also simulated by [69]. They presented energy and exergy analyses for the cycle. The impact of climate difference on the design and sizing of CCHP systems is studied by [70]. They used the MRM method to size the engine of the CCHP cycle in five different climates of Iran. They also used multicriteria decision-making/sizing methods for selecting the prime mover type/size [71] and [72]. Multiobjective optimization and risk analyses are applied to a residential CCHP system in [73]. They used a genetic algorithm for the optimization. Life-cycle assessment (LCA) is also applied to CCHP systems to study the benefits of CCHP systems with respect to SCHP systems by [74]. Matrix approach modeling is also used for sizing the prime mover of a CCHP system by [75].

The impact of off-design equipment on the performance of a CCHP system is presented by [76]. The target function to be minimized in this sizing technique is the total annual cost. They compared the results for two cases of constant efficiencies and off-design characteristics of equipment.

The latest research about CCHP systems was studied in the literature review. The purpose of this review was to recognize designed CCHP cycles, corresponding technologies used in cycles, decision-making methods, design criteria, optimization algorithms, strategies used for energy management, and equipment sizing techniques. In the following chapter the basic CCHP cycles will be discussed in more detail. Technical, environmental, and economic characteristics of the corresponding technologies will be presented to create a strong foundation for the design and optimization of the CCHP cycles and the decision-making required.

References

[1] Oliveira, A.C., Afonso, C., Matos, J., Riat, S., Nguyen, M., Doherty, P., 2002. A Combined Heat And Power System for Buildings Driven by Solar Energy and Gas. Applied Thermal Engineering 22, 587–593.

[2] Pilavachi, P.A., 2002. Mini- And Micro-Gas Turbines for Combined Heat and Power. Applied Thermal Engineering 22, 2003–2014.

[3] Cardona, E., Piacentino, A., 2003. A Methodology for Sizing a Trigeneration Plant in Mediterranean Areas. Applied Thermal Engineering 23, 1665–1680.

[4] Kong, X.Q., Wang, R.Z., Huang, X.H., 2004. Energy Efficiency and Economic Feasibility of CCHP Driven by Stirling Engine. Energy Conversion and Management 45, 1433–1442.

[5] Bruno, J.C., Valero, A., Coronas, A., 2005. Performance Analysis of Combined Microgas Turbines and Gas Fired Water/LiBr Absorption Chillers with Post-Combustion. Applied Thermal Engineering 25, 87–99.

[6] Invernizzi, C., Iora, P., 2005. Heat Recovery from a Micro-Gas Turbine by Vapour Jet Refrigeration Systems. Applied Thermal Engineering 25, 1233–1246.

[7] Katsigiannis, P.A., Papadopoulos, D.P., 2005. A General Technoeconomic and Environmental Procedure for Assessment of Small-Scale Cogeneration Scheme Installations: Application to a Local Industry Operating in Thrace, Greece, Using Microturbines. Energy Conversion and Management 46, 3150–3174.

[8] Wu, D.W., Wang, R.Z., 2006. Combined Cooling, Heating and Power: A Review. Progress in Energy and Combustion Science 32, 459–495.

[9] Onovwiona, H.I., Ugursal, V.I., 2006. Residential Cogeneration Systems: Review of the Current Technology. Renewable and Sustainable Energy Reviews 10, 389–431.

[10] Yagoub, W., Doherty, P., Riffat, S.B., 2006. Solar Energy-Gas Driven Micro-CHP System for an Office Building. Applied Thermal Engineering 26, 1604–1610.

[11] Biezma, M.V., San Cristbal, J.R., 2006. Investment Criteria for the Selection of Cogeneration Plants—A State of the Art Review. Applied Thermal Engineering 26, 583–588.

[12] Invernizzi, C., Iora, P., Silva, P., 2007. Bottoming Micro-Rankine Cycles for Micro-Gas Turbines. Applied Thermal Engineering 27, 100–110.

[13] Ehyaei, M.A., Bahadori, M.N., 2007. Selection of Micro Turbines to Meet Electrical and Thermal Energy Needs of Residential Buildings in Iran. Energy and Buildings 39, 1227–1234.

[14] Godefroy, J., Boukhanouf, R., Riffat, S., 2007. Design, Testing and Mathematical Modelling of A Small-Scale CHP and Cooling System (Small CHP-Ejector Trigeneration). Applied Thermal Engineering 27, 68–77.

[15] Paepe, M.D., Mertens, D., 2007. Combined Heat and Power in a Liberalised Energy Market. Energy Conversion and Management 48, 2542–2555.

[16] Huangfu, Y., Wu, J.Y., Wang, R.Z., Kong, X.Q., Wei, B.H., 2007. Evaluation and Analysis of Novel Micro-Scale Combined Cooling, Heating and Power (MCCHP) System. Energy Conversion and Management 48, 1703–1709.

[17] Li, C.Z., Shi, Y.M., Huang, X.H., 2008. Sensitivity Analysis of Energy Demands on Performance of CCHP System. Energy Conversion and Management 49, 3491–3497.

[18] Hao, X., Yang, H., Zhang, G., 2008. Trigeneration: A New Way for Landfill Gas Utilization and Its Feasibility in Hong Kong. Energy Policy 36, 3662–3673.

[19] Wang, J.-J., Jing, Y.-Y., Zhang, C.-F., Shi, G.-H., Zhang, X.-T., 2008. A Fuzzy Multi-Criteria Decision-Making Model for Trigeneration System. Energy Policy 36, 3823–3832.

[20] Wang, J.-J., Jing, Y.-Y., Zhang, C.-F., Zhang, X.-T., Shi, G.-H., 2008. Integrated Evaluation of Distributed Triple-Generation Systems Using Improved Grey Incidence Approach. Energy 33, 1427–1437.

[21] Mago, P.J., Chamra, L.M., 2009. Analysis and Optimization of CCHP Systems Based on Energy, Economical, and Environmental Considerations. Energy and Buildings 41, 1099–1106.

[22] Wang, J.-J., Jing, Y.-Y., Zhang, C.-F., 2009. Optimization of Capacity and Operation for CCHP System by Genetic Algorithm. Applied Energy 87, 1325–1335.

[23] Cho, H., Mago, P.J., Luck, R., Chamra, L.M., 2009. Evaluation of CCHP Systems Performance Based on Operational Cost, Primary Energy Consumption, and Carbon Dioxide Emission by Utilizing an Optimal Operation Scheme. Applied Energy 86, 2540–2549.

[24] Sugiartha, N., Tassou, S.A., Chaer, I., Marriott, D., 2009. Trigeneration in Food Retail: An Energetic, Economic and Environmental Evaluation for a Supermarket Application. Applied Thermal Engineering 29, 2624–2632.

[25] Lozano, M.A., Ramos, J.C., Carvalho, M., Serra, L.M., 2009. Structure Optimization of Energy Supply Systems in Tertiary Sector Buildings. Energy and Buildings 41, 1063–1075.

[26] Lozano, M.A., Carvalho, M., Serra, L.M., 2009. Operational Strategy and Marginal Costs in Simple Trigeneration Systems. Energy 34, 2001–2008.

[27] Sanaye, S., Ardali, M.R., 2009. Estimating the Power and Number of Microturbines in Small-Scale Combined Heat and Power Systems. Applied Energy 86, 895–903.

[28] Magnani, F.S., Melo, N.R.D., 2009. Use of the Task Configuration System (TCS) for the Design and On-Line Optimization of Power Plants Operating with Variable Loads. Applied Thermal Engineering 29, 455–461.

[29] Lai, S.M., Hui, C.W., 2009. Feasibility and Flexibility for a Trigeneration System. Energy 34, 1693–1704.

[30] Lai, S.M., Hui, C.W., 2009. Integration of Trigeneration System and Thermal Storage under Demand Uncertainties. Applied Energy 87, 2868–2880.

[31] Bruno, J.C., Ortega-L Pez, V., Coronas, A., 2009. Integration of Absorption Cooling Systems into Micro Gas Turbine Trigeneration Systems Using Biogas: Case Study of a Sewage Treatment Plant. Applied Energy 86, 837–847.

[32] Wang, J., Dai, Y., Gao, L., Ma, S., 2009. A New Combined Cooling, Heating and Power System Driven by Solar Energy. Renewable Energy 34, 2780–2788.

[33] Dai, Y., Wang, J., Gao, L., 2009. Parametric Optimization and Comparative Study of Organic Rankine Cycle (ORC) for Low Grade Waste Heat Recovery. Energy Conversion and Management 50, 576–582.

[34] J. Facão and A.C. Oliveira, "Analysis of Energetic, Design and Operational Criteria When Choosing an Adequate Working Fluid for Small ORC Systems," *Proceeding of the ASME 2009 International Mechanical Engineering Congress And Exposition* (IMECE 2009) November 13-19, Lake Buena Vista, Florida, USA.

[35] Wang, J., (John) Zhai, Z., Jing, Y., Zhang, C., 2010. Particle Swarm Optimization for Redundant Building Cooling Heating and Power System. Applied Energy 87, 3668–3679.

[36] Mago, P.J., Hueffed, A.K., 2010. Evaluation of a Turbine Driven CCHP System for Large Office Buildings under Different Operating Strategies. Energy and Buildings 42, 1628–1636.

[37] Mago, P.J., Hueffed, A., Chamra, L.M., 2010. Analysis and Optimization of the Use of CHP–ORC Systems for Small Commercial Buildings. Energy and Buildings 42, 1491–1498.

[38] Ren, H., Zhou, W., Nakagami, K., Gao, W., 2010. Integrated Design and Evaluation of Biomass Energy System Taking into Consideration Demand Side Characteristics. Energy 35, 2210–2222.

[39] Alanne, K., Serholm, N., Sirén, K., B-Morrison, I., 2010. Techno-Economic Assessment and Optimization of Stirling Engine Micro-Cogeneration Systems in Residential Buildings. Energy Conversion and Management 51, 2635–2646.

[40] Srinivasan, K.K., Mago, P.J., Krishnan, S.R., 2010. Analysis of Exhaust Waste Heat Recovery from a Dual Fuel Low Temperature Combustion Engine Using an Organic Rankine Cycle. Energy 35, 2387–2399.

[41] Ameri, M., Behbahaninia, A., Tanha, A.A., 2010. Thermodynamic Analysis of a Tri-Generation System Based on Micro-Gas Turbine with a Steam Ejector Refrigeration System. Energy 35, 2203–2209.

[42] Tichi, S.G., Ardehali, M.M., Nazari, M.E., 2010. Examination of Energy Price Policies in Iran for Optimal Configuration of CHP and CCHP Systems Based on Particle Swarm Optimization Algorithm. Energy Policy 38, 6240–6250.

[43] Li, C.-Z., Shi, Y.-M., Liu, S., Zheng, Z.-L., Liu, Y.-C., 2010. Uncertain Programming of Building Cooling Heating and Power (BCHP) System Based on Monte-Carlo Method. Energy and Buildings 42, 1369–1375.

[44] Wakui, T., Yokoyama, R., 2010. Optimal Sizing of Residential Gas Engine Cogeneration System for Power Interchange Operation from Energy-Saving Viewpoint. Energy 36, 3816–3824.

[45] MartiNez-Lera, S., Ballester, J., 2010. A Novel Method for the Design of CHCP (Combined Heat, Cooling and Power) Systems for Buildings. Energy 35, 2972–2984.

[46] Díaz, P.R., Benito, Y.R., Parise, J.A.R., 2010. Thermoeconomic Assessment of a Multi-Engine, Multi-Heat-Pump CCHP (Combined Cooling, Heating and Power Generation) System – A Case Study. Energy 35, 3540–3550.

[47] Angrisani, G., Minichiello, F., Roselli, C., Sasso, M., 2010. Desiccant Hvac System Driven by a Micro-CHP: Experimental Analysis. Energy and Buildings 42, 2028–2035.

[48] Sun, Z.-G., Xie, N.-L., 2010. Experimental Studying of a Small Combined Cold and Power System Driven by a Micro Gas Turbine. Applied Thermal Engineering 30, 1242–1246.

[49] Luo, H., Wangb, R., Dai, Y., 2010. The Effects of Operation Parameter on the Performance of a Solar-Powered Adsorption Chiller. Applied Energy 87, 3018–3022.

[50] El-Sharkawy, I.I., 2010. On the Linear Driving Force Approximation for Adsorption Cooling Applications. International Journal of Refrigeration. doi: 10.1016/J. Ijrefrig.2010.12.006.

[51] Deng, J., Wang, R.Z., Han, G.Y., 2010. A Review of Thermally Activated Cooling Technologies for Combined Cooling, Heating and Power Systems. Progress in Energy and Combustion Science 37, 172–203.

[52] Angrisani, G., Capozzoli, A., Minichiello, F., Roselli, C., Sasso, M., 2011. Desiccant Wheel Regenerated by Thermal Energy from a Micro Cogenerator: Experimental Assessment of the Performances. Applied Energy 88, 1354–1365.

[53] Hassan, H.Z., Mohamada, A.A., Bennacer, R., 2011. Simulation of an Adsorption Solar Cooling System. Energy 36, 530–537.

[54] Fumo, N., Mago, P.J., Jacobs, K., 2011. Design Considerations for Combined Cooling, Heating, and Power Systems at Altitude. Energy Conversion and Management 52, 1459–1469.

[55] Zhang, C., Yang, M., Lu, M., Shan, Y., Zhu, J., 2011. Experimental Research on LiBr Refrigeration Heat Pump System Applied in CCHP System. Applied Thermal Engineering 31, 3706–3712.

[56] Abdollahi, Gh., Meratizaman, M., 2011. Multi-objective Approach in Thermoenvironomic Optimization of a Small-Scale Distributed CCHP System with Risk Analysis. Energy and Buildings 43, 3144–3153.

[57] Wang, J.-J., Jing, Y.-Y., Zhang, C.-F., (John) Zhai, Z., 2011. Performance Comparison of Combined Cooling Heating and Power System in Different Operation Modes. Applied Energy 88, 4621–4631.

[58] Wang, Y., Huang, Y., Chiremba, E., Roskilly, A.P., Hewitt, N., Ding, Y., Wu, D., Yu, H., Chen, X., Li, Y., Huang, J., Wang, R., Wu, J., Xia, Z., Tanf, C., 2011. An Investigation of a Household Size Trigeneration Running with Hydrogen. Applied Energy 88, 2176–2182.

[59] Parise, J.A.R., Martínez, L.C.C., Marques, R.P., Mena, J.B., Vargas, J.V.C., 2011. A Study of the Thermodynamic Performance and CO_2 Emissions of a Vapour Compression Bio-trigeneration System. Applied Thermal Engineering 31, 1411–1420.

[60] Ehyaei, M.A., Ahmadi, P., Atabi, F., Heibati, M.R., Khorshidvand, M., 2012. Feasibility Study of Applying Internal Combustion Engines in Residential Buildings by Exergy, Economic and Environmental Analysis. Energy and Buildings 55, 405–413.

[61] Sheikhi, A., Ranjbar, A.M., Oraee, H., 2012. Financial Analysis and Optimal Size and Operation for a Multicarrier Energy System. Energy and Buildings 48, 71–78.

[62] Nosrat, A., Pearce, J.M., 2011. Dispatch Strategy and Model for Hybrid Photovoltaic and Trigeneration Power Systems. Applied Energy 88, 3270–3276.

[63] Liu, M., Shi, Y., Fang, F., 2012. A New Operation Strategy for CCHP Systems with Hybrid Chillers. Applied Energy 95, 164–173.

[64] Liu, M., Shi, Y., Fang, F., 2012. Optimal Power Flow and PGU capacity of CCHP Systems Using a Matrix Modeling Approach. Applied Energy 102, 794–802.

[65] Wang, J., Zhao, P., Niu, X., Dai, Y., 2012. Parametric Analysis of a New Combined Cooling, Heating and Power System with Transcritical CO_2 Driven by Solar Energy. Applied Energy 94, 58–64.

[66] Gu, Q., Ren, H., Gao, W., Ren, J., 2012. Integrated Assessment of Combined Cooling Heating and Power Systems under Different Design and Management Options for Residential Buildings in Shanghai. Energy and Buildings 51, 143–152.

[67] Jing, Y.-Y., Bai, H., Wang, J.-J., 2012. A Fuzzy Multi-criteria Decision-Making Model for CCHP Systems Driven by Different Energy Sources. Energy Policy 42, 286–296.

[68] Mago, P.J., Luck, R., 2012. Evaluation of the Potential Use of a Combined Micro-Turbine Organic Rankine cycle for Different Geographic Locations. Applied Energy 102, 1324–1333.

[69] Chen, X.P., Wang, Y.D., Yu, H.D., Wu, D.W., Li, Yapeng, Roskilly, A.P., 2012. A Domestic CHP System with Hybrid Electrical Energy Storage. Energy and Buildings 55, 361–368.

[70] Ebrahimi, M., Keshavarz, A., Jamali, A., 2012. Energy and Exergy Analyses of a Micro-Steam CCHP Cycle for a Residential Building. Energy and Buildings 45, 202–210.

[71] Ebrahimi, M., Keshavarz, A., 2012. Climate Impact on the Prime Mover Size and Design of a CCHP System for the Residential Building. Energy and Buildings 54, 283–289.

[72] Ebrahimi, M., Keshavarz, A., 2012. Prime Mover Selection for a Residential Micro-CCHP by Using Two Multi-Criteria Decision-Making Methods. Energy and Buildings 55, 322–331.

[73] Ebrahimi, M., Keshavarz, A., 2013. Sizing the Prime Mover of a Residential Micro-CCHP System by Multi-Criteria Sizing Method for Different Climates. Energy 54, 291–301.

[74] Sayyaadi, H., Abdollahi, G., 2013. Application of the Multi-Objective Optimization and Risk Analysis for the Sizing of a Residential Small-Scale CCHP System. Energy and Buildings 60, 330–344.

[75] Maraver, D., Sin, A., Sebastian, F., Royo, J., 2013. Environemental Assessment of CCHP (Combined Cooling Heating and Power) Systems Based on Biomass Combustion in Comparison to Conventional Generation. Energy 57, 17–23.

[76] Liu, M., Shi, Y., Fang, F., 2013. Optimal Power Flow and PGU Capacity of CCHP Systems Using a Matrix Modeling Approach. Applied Energy 102, 794–802.

[77] Giffin, P.K., 2013. Performance and Cost Results from a DOE Micro-CHP Demonstration Facility at Mississippi State University. Energy Conversion and Management 65, 364–371.

CCHP Technology

2

2.1   Introduction

By combining different types of prime movers with different heating and cooling systems, many CCHP systems can be designed hypothetically. According to the previous chapter, the prime movers that can be used in CCHP systems include different types of industrial steam turbines (ST), industrial gas turbines (GT), reciprocating internal combustion engines[1] (IC), micro-gas turbines (MGT), micro-steam turbines (MST), different types of Stirling engines (STR) and fuel cells. In addition, the cooling systems that are common in CCHP systems include absorption chillers, adsorption chillers, desiccant dehumidifiers, and ejector cooling systems. Moreover electrical compression chillers are used in parallel with other cooling systems. Other equipment includes different boilers, heaters, heat exchangers, pumps, generators, etc. Because of the differences in operation, price, and environmental effects of prime movers, CCHP systems also operate differently. To help in becoming familiar with the principals of operation of CCHP systems, in the following we present the basic CCHP cycles.

2.2   Basic CCHP Cycles

2.2.1   CCHP Based on Industrial Steam Turbines (ST)

Industrial steam turbines use saturated or superheated pressurized steam to rotate the rotor of the steam turbine. Steam can be produced by burning fossil fuels such as natural gas, and releasing their chemical energy to heat the high-pressure liquid in the boiler tubes (water wall tubes, risers, down-comers, superheaters, and economizers). Steam can also be produced by a heat recovery steam generator (HRSG) placed, for example, at the exhaust of a gas turbine. Some steam turbines also can use the low-pressure steam of some processes to produce power. The thermal energy and potential energy of steam is converted to kinetic energy due to steam expansion in the stationary nozzle buckets (called stators) of the steam turbine; steam jets containing high kinetic energy produce mechanical energy when the rotor rotates due to these jets striking the rotor buckets. This mechanical energy can be used to produce electricity by coupling the steam turbine rotor to a generator. Steam can remain in the steam turbine and produce mechanical energy until its pressure reaches the condenser pressure or downstream pressure of the turbine. Another parameter that restricts the steam residue time in the steam turbine is the steam quality and liquid water content in the steam when it reaches the last stages of the steam turbine. The steam temperature, pressure, and quality are important when it leaves the turbine if it is supposed to be used for cooling and heating purposes in CCHP systems.

---

[1]The spark ignition and compression types of IC.

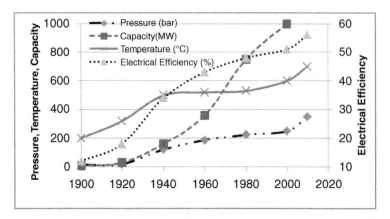

Figure 2.1 Basic characteristics of industrial steam turbines [1].

Figure 2.1 shows the basic characteristics of steam turbines such as operating pressure, temperature, capacity, and thermal efficiency produced by the Siemens Company [1] in the last century. The steam turbines presented in this figure are simple steam turbines without reheating (from 1900 to 1920), steam turbines with reheating (from 1920 to 1960) and supercritical steam turbines (from 1960).

Industrial steam turbines are categorized based on application, construction, and bucket row type. A diagram of these classifications is given in Figure 2.2.

According to the applications summarized in Figure 2.2, steam turbines can be classified into six types. *Saturated steam turbines* work with saturated steam; additional superheating equipment is not needed. They can be utilized to produce power from low quality steam, ram pumps, blowers, etc. They may be used in organic Rankine cycles (ORC) as well, because some of the refrigerants become superheated automatically as they enter the turbine in saturated condition. *Low-pressure steam turbines* make use of the low-pressure steam of processes or other turbines to produce electricity, turn water pumps, operate blowers, etc. The steam inlet pressure of these turbines is usually smaller than 1 MPa. *Condensing steam turbines* bring the exhaust pressure

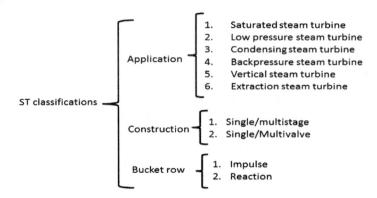

Figure 2.2 Steam turbine classifications.

of the turbine to below the atmospheric pressure. Therefore they are equipped with a condensing system that works under the atmospheric pressure. Sealing of the condenser against entering ambient air is critical. This type of steam turbine is not proper for CCHP purposes, because the exhaust has very low pressure and temperature.

Backpressure steam turbines have relatively high-pressure exhaust steam; therefore the exhaust can be used for running low-pressure steam turbines, condensing steam turbines, or ORC. The exhaust of these turbines can also be used for heating or cooling purposes (CCHP). Vertical steam turbines use low-pressure steam to run compressors, blowers, and pumps in vertical position. *Extraction/induction steam turbines* have extraction/induction lines in the intermediate stages. The extraction line can be used for CCHP purposes. The induction line also can be used for CCHP purposes, because induction steam can be used for cooling or heating before induction into the turbine.

As discussed above, among different types of steam turbines only the backpressure and extraction/induction steam turbines can be used for CCHP purposes. Figures 2.3A and B show the application of backpressure and extraction/induction steam turbines in CCHP systems.

Industrial steam turbines usually can produce several MWs of electricity; therefore they can be used for large-scale CCHP systems that are especially proper for hospitals, large commercial, or residential complexes [2]. CCHP systems are classified into large scale (greater than 1 MW), small scale (smaller than 1 MW), mini (smaller than 500 kW) and micro (smaller than 20 kW) for [3].

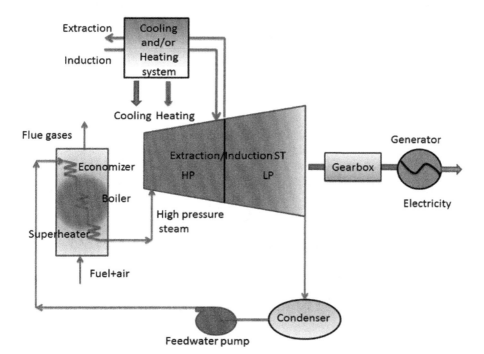

Figure 2.3A CCHP system based on extraction/induction steam turbine.

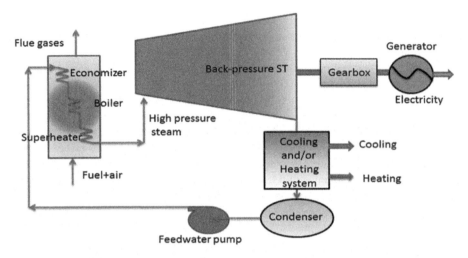

Figure 2.3B CCHP system based on backpressure steam turbine.

Steam turbines work in a wide range of steam pressures. The steam pressure can reach as high as 3500 psig (241.32 bar (g)) at the inlet and as low as 0.5 psia (0.034 bar (a)) at the exit of the steam turbine [4]. Steam turbines can work with different gaseous, liquid, and solid fuels. They have a very long lifetime (more than 50 years) if operated and maintained properly. Controlling high and low water level is very important to avoid harmful phenomena such as carryover and overheating in the turbine and boiler tubes. Also, chemical treatment of water is extremely important to avoid scale formation and corrosion in the boiler tubes, drum, and steam turbine. In addition steam quality in the inlet and outlet of the turbine should be controlled to avoid erosion of turbine blades. Moreover, combustion should be controlled to keep the boiler tubes clean from outside elements. Every type of deposit on the outside surface of tubes can result in overheating and failure of tubes.

Steam turbines are constructed in sizes from about 100 kW to more than 250 MW. Since steam turbines are designed to provide base loads, they usually work constantly for a long time and steam temperature experiences small changes during operation. The startup time of steam turbines is long; large steam turbines may take 24 hours or more to start up. In Table 2.1, some characteristics of three sizes of steam turbine from TurboSteam, Inc. are compared and presented [4].

In Table 2.1 the electric heat rate, net heat rate, effective electrical efficiency (*EEE*), and recovered heat to inlet fuel energy ratio (RFR) are defined as below:

$$Electric\ heat\ rate = \frac{F}{E_{PM}} \tag{2-1}$$

$$Net\ heat\ rate = \frac{F - Q_{rec}/\eta_b}{E_{PM}} \tag{2-2}$$

Table 2.1 **Steam Turbine Characteristics [4]**

Steam Turbine Characteristics			
E_{nom} (kW)	500	3000	15000
Turbine type	Backpressure	Backpressure	Backpressure
Typical application	Chemical plants	Paper mill	Paper mill
Equipment cost[*] (2008 $/kW)	657	278	252
Total installed cost (2008 $/kW)	1117	475	429
Turbine isentropic efficiency (%)	50	70	80
Generator/gearbox efficiency (%)	94	94	97
Steam flow (lbs/hr)	21500	126000	450000
Steam flow (kg/hr)	9752.2	57152.6	204116.6
Inlet pressure (bar(g))	34.5	41.4	48.3
Outlet pressure (bar(g))	3.4	10.3	10.3
Inlet temperature (°C)	287.8	301.7	343.3
Outlet temperature (°C)	147.8	185.5	185.5
CHP Characteristics			
Boiler efficiency (%)	80	80	80
CHP electric efficiency (%)	6.4	6.9	9.3
F (MMBtu/hr)	26.7	147.4	549.0
F (kW)	7818.0	43160.3	160752.9
Steam to process (kW)	5740	31352	113291
Steam to process (MMBtu/hr)	19.6	107.0	386.6
η_o(%), HHV	79.6	79.5	79.7
PHR	0.09	0.10	0.13
Net heat rate (Btu/kWh)	4515	4568	4388
EEE (%), HHV	75.6	75.1	77.8
RFR (%)	73	72	70

* Equipment cost does not include the boiler and steam system costs. It includes turbine, gearbox, generator, control, and switchgear.

$$EEE = \frac{E_{PM}}{F - Q_{rec}/\eta_b} \qquad (2\text{-}3)$$

$$RFR = \frac{Q_{rec}}{F} \qquad (2\text{-}4)$$

where η_b is a typical boiler efficiency that produces the same amount of heat as Q_{rec}. In the above calculations $\eta_b = 80\%$ is assumed.

Emission of steam turbines is related to the emission of the boiler. Depending on the fuel type, the magnitude and type of pollution is very different. In Table 2.2, the emissions of fuel oil and natural gas in the boiler of the steam turbines mentioned in Table 2.1 are presented.

Table 2.2 Emissions of Fuel Oil and Natural Gas in the Boiler of Steam Turbines [4]

Boiler Fuel	500 kW			3 MW and 15 MW		
	NO_x	CO	*Particulate matter*	NO_x	CO	*Particulate matter*
Fuel oil (lbs/MMBtu)	0.15-0.37	0.03	0.01-0.08	0.07-0.31	0.3	0.01-0.08
Natural gas (lbs/MMBtu)	0.03-0.1	0.08	-	0.1-0.28	0.08	-

2.2.2   CCHP System Based on Industrial Gas Turbine (GT)

Industrial gas turbines use ambient air as the main working fluid of the cycle. An industrial land gas turbine has six main components: an air-intake system, which is also called a filter house; a compressor that can be axial, centrifugal, or a combination of axial and centrifugal; a combustion chamber; a turbine; an exhaust; and a starter (Figure 2.4).

The starter turns the compressor to draw in fresh ambient air through the filter house. The starter will be disconnected from the compressor when the turbine produces enough power to turn the compressor.

The filter house produces clean air with proper temperature and humidity and noise. A filter house has some filter elements to clean the air. A self-cleaning system produces an inverse jet pulse of air into each filter element when the pressure drop

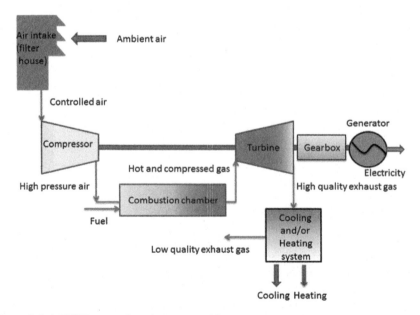

Figure 2.4 A CCHP system based on a gas turbine.

across the filters reaches a predefined set point. The fog system decreases the air temperature during hot days through injection of demineralized water mist into the clean air using some nozzles; some louvers prevent the liquid water droplets from entering the compressor. An anti-icing system injects hot air extracted from the last stages of the compressor to the filter house to avoid the impact of freezing air on the filters and compressor inlet. Air flowing through the filter house produces noise; to keep the noise down some silencers are placed inside the filter house just before entering the compressor.

In the second step air enters the compressor, its pressure increases while passing through the compressor stages, and reaches the maximum possible pressure at the last stage of the compressor. The flow rate and direction of air entering the compressor are controlled using variable inlet guide vanes (VIGVs). VIGVs are also helpful to avoid stall on the compressor blades. The compressor also has some bleed valves to avoid surge at startup and shutdown. Bleed valves usually vent air into the exhaust or bypass it to the compressor inlet. Air is also extracted from the compressor casing to cool down the hot stages of the turbine (usually the first and second stages). The anti-icing system also makes use of air extraction from the compressor casing.

In the third step pressurized air enters the combustion chamber and after burning with fuel, its temperature also increases. The rate at which fuel enters the combustion chamber is controlled by the governor system according to the rotor speed, air flow rate, maximum pressure, compressor inlet, exit temperature, and exhaust temperature. Igniters generate flame inside the combustion chamber and flame detectors check the flame condition. To avoid overheating and melting of combustion chamber components, extra air enters the combustion chamber. Less than 20% of air participates in the combustion process; the remaining air is consumed for recirculation, cooling, and dilution purposes inside the combustion chamber.

In the fourth step, combustion products and extra hot air enter the turbine with the allowable turbine design temperature and pressure. The turbine must convert the thermal and potential energy of entering gases to the rotational motion of the rotor as much as possible. For this purpose the thermal and potential energy of gas are converted to kinetic energy in the stationary nozzles, and then this energy strikes the rotor blades, converting the kinetic energy to mechanical rotation of the turbine rotor. Relatively cool air extracted from the last stages of the compressor is channeled to cool the stationary and rotary blades of some stages of the turbine internally and externally using convection and film cooling mechanisms. While the gas is expanding through the turbine, all of its potential energy and a portion of its thermal energy are converted to shaft rotation. Shaft rotation can be converted to electricity by coupling it with a generator. The portion of thermal energy that was not converted to rotor rotation exits from the turbine and enters the exhaust. When a gas turbine is supposed to be used as the prime mover of a CCHP system, the heat of exhaust gases can be recovered to produce cooling or heating. The quality of recoverable heat from the prime movers has a significant impact on the feasibility and overall efficiency of CCHP systems. A schematic of a CCHP system based on a gas turbine is shown in Figure 2.4.

The exhaust gas of a gas turbine has a temperature as high as 540 °C, which makes this technology very suitable for CCHP systems. They work with different kind of fuels such as natural gas, landfill gas, synthetic gas, and liquid fuels such as diesel. Gas turbines are reliable in operation and maintenance; their overhaul period is 25,000 to 50,000 hours of operation. In addition, gas turbine capacity ranges from 500 kW to about 250 MW, which means they can be used for large- and small-scale CCHP applications. Using DLE (dry low emission) technology in the combustor of gas turbines has decreased NO_x and CO production to about 25 ppm and 10-50 ppm, respectively [4]. Gas turbines have poor partial load efficiency and power production because they reduce power generation by decreasing fuel injection; consequently the maximum temperature decreases as well. Decreasing the maximum temperature decreases the efficiency significantly. In Table 2.3, five gas turbines with different capacities are

Table 2.3 Comparison of Technical and Economic Characteristics of Gas Turbines [4]

Cost and Performance Characteristics					
E_{PM} (kWe)	1150	5475	10239	25000	40000
E_{nom} (kWe)	1000	5000	10000	25000	40000
Basic installed cost (2007 $/kW)	3324	1314	1298	1097	972
Complete installation cost (2007 $/kW)	5221	2210	1965	1516	1290
Electric heat rate HHV (Btu/kWh)	16087	12274	12003	9944	9220
$\eta_e(\%)$, HHV	21.27	27.72	28.44	34.30	37.00
F (MMBtu/hr)	18.5	67.2	122.9	248.6	368.8
F (kW)	5417	19677	35986	72793	107989
Required fuel gas pressure (psig)	82.6	216	317.6	340	435
CHP Characteristics					
Exhaust flow (1000lb/hr)	51.4	170.8	328.2	571	954
GT exhaust temperature (°C)	511	516	491	510	457
HRSG exhaust temperature (°C)	154	153	161	138	138
Q_{rec} (MMBtu/hr)	8.31	28.26	49.10	90.34	129.27
Q_{rec} (1000lbs/hr)	8.26	28.09	48.80	89.8	128.5
Q_{rec} (kW)	2435	8279	14385	26469	37876
$\eta_o(\%)$, HHV	66.3	69.8	68.4	70.7	72.1
PHR	0.47	0.66	0.71	0.94	1.06
Form of recovered heat	Saturated Steam	Saturated Steam	Saturated Steam	Saturated Steam	Saturated Steam
RFR (%)	45	42	40	36	35
Net heat rate (Btus/kWh)	7013	5839	6007	5427	5180
EEE (%)	49	58	57	63	66

compared from technical and economic points of view. The exhaust gas of these turbines is used to produce saturated steam in a HRSG. The data presented in the Table 2.3 are for gas turbine sizes of 1150, 5475, 10,239, 25,000, and 40,000 kW; they represent the Solar Turbines Saturn, Solar Turbines Taurus, Solar Turbines Mars, GE LM2500, and GE LM6000, respectively [4]. The term *basic installed cost* in the table is based on the CHP system producing 150 psig saturated steam with an unfired HRSG, no gas compression, no building, and no exhaust treatment in an uncompli-cated installation at a customer site [4]. In addition the term *complete installation cost* refers to an installation at an existing customer site with access constraints, completed electrical, fuel, water, steam connections requiring added engineering, construction costs, and gas compression from 150 psig. The cost also includes building, SCR,[2] CO catalyst,[3] and CEMS[4] [4].

Curve fitting the important technical data presented in the Table 2.3 results in some useful correlations between GT size and other parameters such as electrical efficiency, overall efficiency, fuel consumption, and recoverable heat from the exhaust:

$$\eta_e(\%) = 20.794 E_{nom}^{0.1535}, \quad 1.15 \le E_{nom}(MW) \le 40 \tag{2-5}$$

$$F(MW) = 4.8115 E_{nom}^{0.8458}, \quad 1.15 \le E_{nom}(MW) \le 40 \tag{2-6}$$

$$Q_{rec}(MW) = 2.2295 E_{nom}^{0.774}, \quad 1.15 \le E_{nom}(MW) \le 40 \tag{2-7}$$

$$\eta_o(\%) = 0.2078 E_{nom}^{0.1542} + 0.4634 E_{nom}^{-0.0718}, \quad 1.15 \le E_{nom}(MW) \le 40 \tag{2-8}$$

As mentioned previously, gas turbine efficiency in partial load is poor. In compari-son with MGTs and reciprocating engines, gas turbines are less flexible in efficiently operating in partial load. This is due to lowering the gas turbine inlet temperature when it is supposed to work in partial load. Figure 2.5 shows the partial load ef-ficiency of a gas turbine. At full load (100%) the electrical efficiency reaches about 30% while at partial loads of 50% and 10% the electrical efficiency reaches 25% and 12%, respectively.

Gas turbine efficiency and output power are sensitive to the ambient temperature as well. Decreasing ambient temperature increases the air density and consequently the compression work of the compressor decreases. This means more output power and higher efficiency. On the contrary, increasing ambient temperature decreases the air density and as a result the compression work of the compressor increases. This lowers the output power and efficiency as well. To conquer the sensitivity of gas turbines to ambient air temperature variations, a fog system is designed in the air intake of gas turbines. A fog system makes use of evaporative cooling by injecting high-pressure

---

[2]Selective catalytic reduction: a device for NOx reduction by converting NO_x emission to N_2 and H_2O.
[3]CO catalyst converts CO to a less poisonous gas, CO_2.
[4]Continuous Emissions Monitoring Systems.

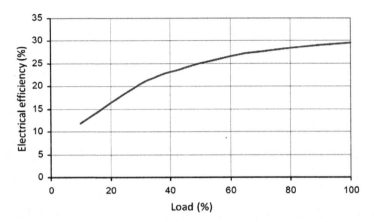

Figure 2.5 Partial load efficiency of gas turbine [4].

demineralized fogged water. A fog system is especially helpful in warm and dry climates. On the other side, cold ambient air temperatures are useful until they produce frost in the air intake filters or compressor inlet. If frosting is possible, the anti-icing system helps to get rid of it. The anti-icing system uses an extraction of warm air from the intermediate or final stages of the compressor to balance the temperature in the air intake and compressor inlet. In Figure 2.6, the impact of ambient temperature from −6.6 °C (20 °F) to 37.7 °C (100 F) is presented [4]. The ISO conditions for the rated power of gas turbines are 1 bar (sea level) and 15 °C.

Altitude above sea level also has an impact on the efficiency and output power of gas turbines. At higher altitudes air density decreases and the compression work of compressor consequently increases. In Figure 2.7 the impact of altitude (0 to 1524 m) on the power output of a gas turbine is demonstrated [4]. As can be seen, at an altitude of 5000 ft (1524m), the output power decreases more than 15%.

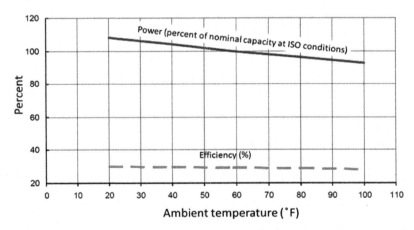

Figure 2.6 Impact of ambient air temperature on the efficiency and output power of a gas turbine [4].

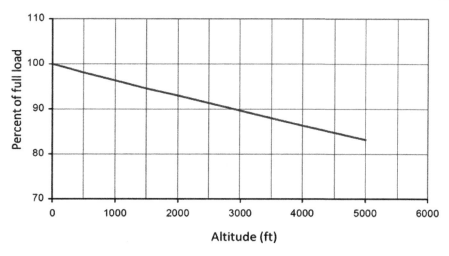

Figure 2.7 Impact of altitude above sea level on the output power of a gas turbine [4].

The operation and maintenance cost for the nominated five gas turbines are presented in the Table 2.4.

The emissions produced by the five sizes of gas turbine presented in Table 2.4 for the untreated exhaust (15% oxygen, no SCR or other emission reducers) are presented in Table 2.5 [4].

Table 2.4 Operation and Maintenance Cost for the Nominated Five Gas Turbines [4]

Operation and Maintenance Costs (I_{OM})					
E_{nom} (kWe)	1000	5000	10000	25000	40000
Variable (service, contract), 2007 $/kWh	0.0060	0.0060	0.0060	0.0040	0.0035
Variable (consumables), 2007 $/kWh	0.0001	0.0001	0.0001	0.0001	0.0001
Fixed, 2007 $/kW-year	40	10	7.5	6	5
Fixed, 2007 $/kWh @ 8000hrs/year	0.005	0.0013	0.0009	0.0008	0.0006
Total O & M costs, 2007 $/kWh	0.0111	0.0074	0.0070	0.0049	0.0042

Table 2.5 Emission Indices for Some Gas Turbines [4]

Emissions Characteristics					
Electricity capacity (kW)	1000	5000	10000	25000	40000
NO_x (ppm)	42	15	15	25	15
NO_x (lb/MWh)	2.43	0.66	0.65	0.9	0.5
CO (ppm)	20	25	25	25	25
CO (lb/MWh)	0.71	0.68	0.66	0.55	0.51
Carbon (lb/MWh)	512	393	383	317	294
CO_2 (lb/MWh)	1877	1440	1404	1163	1079

2.2.3 CCHP Based on Reciprocating Internal Combustion Engine (IC)

The main components of a reciprocating engine include the cylinder, piston, connecting rod, crankshaft, intake valve, and exhaust valve (Figure 2.8). Fuel burns with oxygen inside the cylinder and expands. This gas expansion pushes the piston and connecting rod to create the biggest possible volume between the inside walls of the cylinder and piston head (converting the chemical energy of fuel to a reciprocating motion). This causes the crankshaft to rotate and convert the reciprocating motion to rotational motion. Finally the rotational motion of the crankshaft is converted to electricity using a generator. The CCHP system shown in Figure 2.8 makes use of only the heat of high-quality exhaust gases with a temperature of about 540 °C.

A reciprocating engine has three main sources of recoverable heat: exhaust, jacketing, and lube oil cooling. These three sources have different qualities (different temperatures) and phases (liquid and gas), which make the recovery and use of them more complicated and difficult.

Reciprocating engines basically are classified as spark ignition (Otto cycle) and compression ignition (diesel). The Otto cycles use light fuels such as gasoline or natural gas and combustion starts by using an igniter. Compression ignition engines use heavy fuels such as diesel; the fuel is sprayed into the cylinder and compression increases the temperature of the fuel and air mixture. When the temperature goes above the self-ignition temperature of the fuel, the combustion starts and the chemical energy of the fuel will be released.

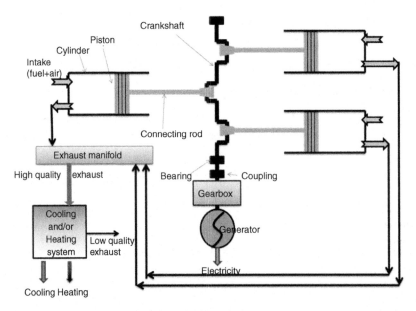

Figure 2.8 A CCHP system based on a reciprocating engine with three-throw crank at 120°.

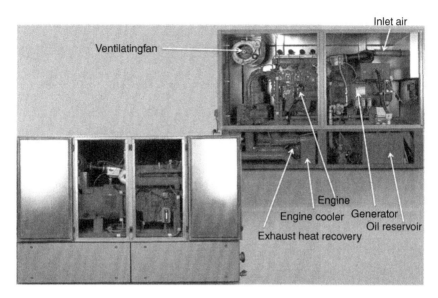

Figure 2.9 The CHP unit of WOLF Company and some main components [5].

Reciprocating internal combustion engine capacity ranges from about 30 kW to hundreds of megawatts. This diversity in capacity makes them suitable for different applications and the demands of large, small, and mini CCHP systems.

WOLF Company has presented CHP systems with capacities from 30 to 360 kWe. Figure 2.9 shows one of these systems.

Important characteristics of the WOLF CHP units such as dimensions, weight, fuel consumption, electricity output, heat recovered, electrical efficiency, PHR, recovered heat to inlet fuel energy ratio (RFR), and overall efficiency are presented in Table 2.6. As can be seen, 51% to 58% of the fuel energy is recovered in different capacities; this huge heat recovery increases the overall efficiency to as high as 85% to 90%. The power to heat ratio increases as the engine size increases (from 0.5 to 0.74).

Curve fitting of the data presented in Table 2.6 shows some correlations between the important parameters, which can be used for mathematical simulations.

Fuel consumption (F) with respect to the nominal size of the engine at full load operation is calculated as follows:

$$F = 2.64E_{nom} + 21.93 \; 30 \leq E_{nom} (kW) \leq 360 \tag{2-9}$$

Recoverable heat from the engine follows the following correlation:

$$Q_{rec} = 1.368E_{nom} + 14.57 \; 30 \leq E_{nom} (kW) \leq 360 \tag{2-10}$$

Table 2.6 **Characteristics of CHP Units Based on Reciprocating Internal Combustion [5]**

Model	Length (m)	Width (m)	Height (m)	Weight (kg)	E_{nom} (Kw)	Q_{rec} (Kw)	Fuel (Kw)	$\eta_e(\%)$	$\eta_o(\%)$	PHR	RFR*
GTK 30	2.4	1.0	1.9	1320	30	60	103	29	87	0.50	0.58
GTK 35	2.4	1.0	1.9	1500	35	60	112	31	85	0.58	0.54
GTK 50	2.4	1.0	1.9	2100	50	79	148	34	87	0.63	0.53
GTK 70	2.8	1.0	1.9	2500	70	109	204	34	88	0.64	0.53
GTK 85M	2.8	1.0	1.9	2500	85	123	240	35	87	0.69	0.51
GTK 140	3.2	1.25	2.15	3600	140	207	392	36	89	0.68	0.53
GTK 200M	3.2	1.25	2.15	3600	198	293	553	36	89	0.68	0.53
GTK 240	3.5	1.5	2.15	4600	236	365	669	35	90	0.65	0.55
GTK 350M	3.4	1.6	2.15	4400	246	344	668	37	88	0.72	0.51
GTK 360M	3.6	1.6	2.15	5900	360	489	955	38	89	0.74	0.51

* Recovered heat to inlet fuel ratio.

Moreover, the nominal electrical efficiency, nominal overall efficiency, and PHR when the engine is operated at full load and all of the recoverable heat is recovered and used can be derived from the following correlations:

$$\eta_e = \frac{E_{nom}}{2.64 E_{nom} + 21.93}$$

$$\eta_o = \frac{2.368 E_{nom} + 14.57}{2.64 E_{nom} + 21.93}, \qquad 30 \le E_{nom}(kW) \le 360 \qquad (2\text{-}11)$$

$$PHR = \frac{E_{nom}}{1.367 E_{nom} + 14.57}$$

The Environmental Protection Agency (EPA) published a report about CHP units with different prime movers such as MGTs, reciprocating internal combustion engines, gas turbines, steam turbines, fuel cells, and so on.

In Table 2.7, some important technical data that were presented in the EPA report are reference for use for design considerations and decision-making.

The input fuel energy to the engine is converted to power and heat. Heat may be lost through exhaust, jacketing, lube oil cooling, and radiating. Heat losses increase in partial load operation, but decrease at full load operation due to higher electrical efficiency. The energy balance of an internal combustion engine is presented in Figure 2.10. As this figure shows, exhaust and jacketing are great sources of energy for recovery. These two sources account for about 50% of the input fuel energy at full load operation. In partial load operation this magnitude increases even more.

Curve fitting of the data presented in Table 2.7, resulting in some correlations between the important criteria.

Table 2.7 Characteristics of Some Gas Engine CHPs [4]

Cost and Performance Characteristics					
E_{nom} (kWe)	100	500	1000	3000	5000
Total installed cost, 2007 $/kW	2210	1940	1640	1130	1130
Electric heat rate, HHV (Btu/kWh)	12000	9866	9760	9493	8758
η_e (%), HHV	28.4	34.6	35	36	39
Engine speed (rpm)	1800	1800	1800	900	720
F (MMBtu/hr)	1.20	4.93	9.76	28.48	43.79
F (kW)	351.37	1443.55	2857.83	8339.24	12822.17
Required fuel gas pressure (psig)	<3	<3	<3	43	65
CHP Characteristics					
Exhaust flow (1000lb/hr)	1.4	6.3	12.1	48.4	67.1
Exhaust temperature (°C)	571	504	487	364	370
Heat recovered from exhaust (MMBtu/hr)	0.28	1.03	1.85	4.94	7.01
Heat recovered from cooling jacket (MMBtu/hr)	0.33	1.13	2.45	4.37	6.28
Heat recovered from lube oil system (MMBtu/hr)	0.00	0.00	0.00	1.22	1.94
Q_{rec} (MMBtu/hr)	0.61	2.16	4.30	10.53	15.23
Q_{rec} (kW)	178.61	632.47	1259.08	3083.29	4459.50
Form of recovered heat	Hot water	Hot water	Hot water	Hot water	Hot water
η_o (%)	79	78	79	73	74
RFR (%)	51	44	44	37	35
PHR	0.56	0.79	0.79	0.97	1.12
Net heat rate (Btus/kWh)	4383	4470	4385	5107	4950
EEE (%)	78	76	78	67	69

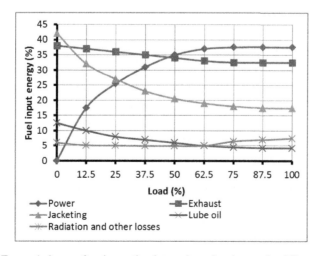

Figure 2.10 Energy balance of reciprocating internal combustion engine [6].

Fuel consumption with respect to the E_{nom} of the engine at fuel load operation can be curve fitted linearly as follows:

$$F = 2.57E_{nom} + 236.17, \ 100 \leq E_{nom}(kW) \leq 5000 \tag{2-12}$$

In addition Q_{rec} follows this correlation:

$$Q_{rec} = 0.87E_{nom} + 247.25, \ 100 \leq E_{nom}(kW) \leq 5000 \tag{2-13}$$

Moreover, the nominal electrical efficiency, nominal overall efficiency, and PHR of the engine when it is operated at full load and all of the recoverable heat is recovered and used can be derived from the above correlations:

$$\begin{aligned} \eta_e &= \frac{E_{nom}}{2.57E_{nom} + 236.17} \\ \eta_o &= \frac{1.87E_{nom} + 247.25}{2.57E_{nom} + 236.17}, \qquad 100 \leq E_{nom}(kW) \leq 5000 \\ PHR &= \frac{E_{nom}}{0.87E_{nom} + 247.25} \end{aligned} \tag{2-14}$$

If the engine is supposed to operate at partial load, knowing the partial load operation efficiency is necessary in simulation and optimization problems.

In Figure 2.11, the ratio of partial load efficiency (η_{PLO}) to the nominal electrical efficiency (η_e) is plotted against the load ratio (E_{PM}/E_{nom}). The curve fitting suggests the following equation for the partial load efficiency:

$$\frac{\eta_{PLO}}{\eta_e}(\%) = \left[-0.0027\left(\frac{E_{PM}}{E_{nom}}(\%)\right)^2 + 0.5693\left(\frac{E_{PM}}{E_{nom}}(\%)\right) + 69.7360 \right],$$
$$30 \leq \frac{E_{PM}}{E_{nom}}(\%) \leq 100 \tag{2-15}$$

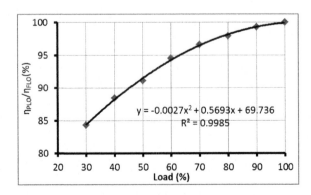

Figure 2.11 Partial load efficiency of internal combustion engine [4].

The ISO conditions in which the performance of reciprocating engines are generally rated are 25 °C and 1 bar. The effects of altitude above sea level and ambient temperature on the efficiency and power output of reciprocating internal combustion engines are not as significant as for gas turbines. As a rule of thumb for every 1000 ft (305 m) of altitude above the first 1000 ft the efficiency and output power degrade by approximately 4%. Increasing ambient temperature decreases engine performance. The output power and efficiency decrease about 1% for every 10 °F (5.5 °C) above 77 °F (25 °C) [4]. The above expressions can be formulated as follows:

$$\eta_e(z) \cong \eta_e \times \left[1 - 0.04\left(\frac{z(m) - 305}{305}\right)\right]$$

$$E_{PM}(z) \cong E_{nom} \times \left[1 - 0.04\left(\frac{z(m) - 305}{305}\right)\right]$$

(2-16)

$$\eta_e(T) \cong \eta_e \times \left[1 - 0.01\left(\frac{T(^\circ C) - 25}{5.5}\right)\right]$$

$$E_{PM}(T) \cong E_{nom} \times \left[1 - 0.01\left(\frac{T^\circ C) - 25}{5.5}\right)\right]$$

(2-17)

In which z is the altitude above the sea level in meter. In Table 2.8, the lb/MWh of different emissions such as CO, CO_2, NO_x, and VOC[5] are reported for CHP units based on reciprocating internal combustion engines. These data are helpful for environmental evaluations.

In addition the maintenance costs (fixed and variable), component costs, including the engine, generator, heat recovery, etc., are presented as rules of thumb and rough calculations in the Tables 2.9 and 2.10. If you intend to create an economic design for CCHP systems, the most updated economic data from the manufacturers and suppliers should be used.

2.2.4 CCHP Based on a Micro-Gas Turbine (MGT)

Micro-gas turbines are packed electricity generators that can use different gas and liquid fuels such as natural gas, gasoline, diesel, biogas, etc. Their single unit capacity

Table 2.8 Combustion and Emission Indices for Reciprocating Internal Combustion Engines [4]

Emissions Characteristics					
Electricity capacity (kW)	100	500	1000	3000	5000
Engine combustion	Rich	Rich	Lean	Lean	Lean
NO_x(lb/MWh)	0.10	0.50	1.49	1.52	1.24
CO (lb/MWh)	0.32	1.87	0.87	0.78	0.75
VOC (lb/MWh)	0.1	0.47	0.38	0.34	0.22
CO_2 (lb/MWh)	1404	1284	1142	1110	1024

[5]Volatile organic compound.

Table 2.9 **Maintenance Costs of Reciprocating Internal Combustion Engines [4]**

Maintenance Costs (I_{OM})					
Capacity (kWe)	100	500	1000	3000	5000
Variable (service, contract), 2007 $/kWh	0.02	0.015	0.012	0.01	0.009
Variable (consumables), 2007 $/kWh	0.00015	0.00015	0.00015	0.00015	0.00015
Fixed, 2007 $/Kw-year	15	7	5	2	1.5
Fixed, 2007 $/kWh @ 8000hrs/year	0.0019	0.0009	0.0006	0.0003	0.0002
Total O & M costs, 2007 $/kWh	0.022	0.016	0.013	0.01	0.009

Table 2.10 **Capital Costs of CHP System Based on Reciprocating Internal Combustion Engines [4]**

Capacity (kWe)	100	500	1000	3000	5000
Costs all in $/kW					
Generator set package	1000	880	760	520	590
Heat recovery	110	240	190	80	50
Interconnect/electrical	260	60	40	30	20
Labor/material	340	300	250	240	250
Project and construction management	200	180	150	90	70
Engineering and fees	200	180	150	90	70
Project contingency	70	60	50	30	30
Project financing (interest during construction)	30	40	50	50	50
Total plant cost	2210	1940	1640	1130	1130

ranges from 15 kW to 250 kW. By paralleling some units, higher capacities can also be achieved for higher demands. Therefore, they can be used for large-, small-, mini-, and micro-scale applications of CCHP systems. The dimensions and weight of a 15 kW MGT unit from a main producer catalog is W × L × H = 0.76 × 1.5× 1.9 (m) and 405 kg, respectively.

The main components of a MGT include a centrifugal compressor, centrifugal turbine, combustion chamber, recuperator, high-frequency generator and its cooler, rectifier, inverter, fuel compressor for low-pressure gaseous fuels, air- or oil-lubricated bearings, and exhaust. The bearings must tolerate a high rotational speed of about 100,000 rpm.

In an MGT air enters the centrifugal compressor as it starts rotation; after that compressed air enters the recuperator for preheating by the exhaust gases coming back from the turbine. This preheating decreases fuel consumption and lowers the exhaust gas temperature from about 500 °C to about 300 °C. Pressurized and pre-heated air enters the combustion chamber and its thermal energy increases due to the chemical energy of fuel released in the combustion process. Gas with the maximum potential and thermal energy enters the centrifugal turbine peripherally and strikes the turbine vanes, causing the turbine impeller to rotate. This rotation is transmitted to the high-frequency generator to produce electricity. The high-frequency AC electricity is rectified to DC in the rectifier and then inverted to 50 or 60 Hz AC. The combustion products after passing through the turbine still have considerable thermal energy. A significant portion of this energy is recovered in the recuperator for preheating compressed air to about 150 °C before entering the combustion chamber. In Figure 2.12, the rotor and a model of the internal parts of a Capstone MGT are presented [7].

The exhaust gases exiting the recuperator can be used for heating or cooling in CCHP applications, because their quality is still high. A schematic of a CCHP system based on a MGT is shown in Figure 2.13.

In order to allow for a better understanding of MGTs, in the following some important technical data are presented. The maximum power output and electrical efficiency of three Capstone MGT models (C30, C65, and C200) in different ambient

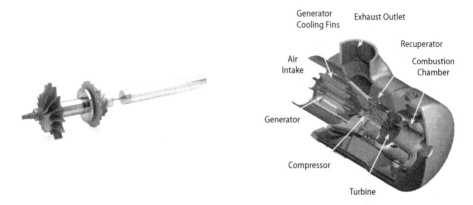

Figure 2.12 Rotor and a model of internal parts of a Capstone MGT [7].

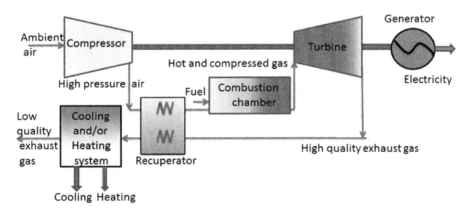

Figure 2.13 A CCHP system based on a micro-gas turbine.

temperatures are presented. These results are at sea level and a relative humidity (RH) of 60%. According to the results, the power generation reduces considerably at temperatures above 65, 60 and 75 °F for the C30, C65, and C200, respectively. Therefore if we are supposed to use an MGT in a CCHP system at non-ISO conditions (59 °F, sea level, and RH of 60%) the impact of changing ambient temperature, altitude, and RH should be included in the analyses. It should be mentioned that Figures 2.14 to 2.16 are modified to improve the quality of the images.

Another piece of technical data that is very important for CCHP systems is the partial load operation efficiency. Due to the large variation of building electrical

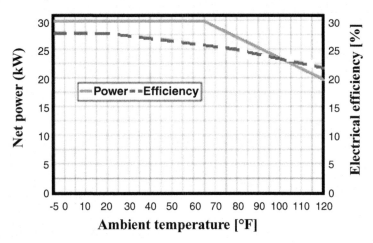

Figure 2.14 Sensitivity of power and electrical efficiency of a C30 to the ambient temperature at sea level with an RH of 60% [7].

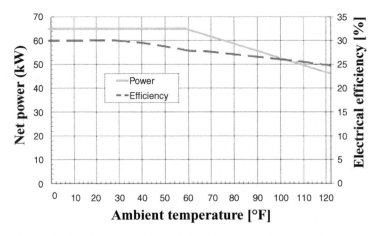

Figure 2.15 Sensitivity of power and electrical efficiency of a C65 low NO$_x$ version to the ambient temperature at sea level with an RH of 60% [7].

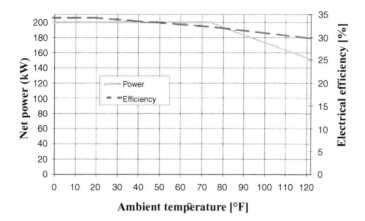

Figure 2.16 Sensitivity of power and electrical efficiency of a C200 to the ambient temperature at sea level with an RH of 60% [7].

demands, it is important to draw a red line for the minimum permissible operation load of the MGT in order to avoid operating the MGT at very low loads and efficiencies. This means that at some working loads it may be more profitable to buy electricity from the grid rather than running the MGT at a very low partial load. For this purpose in Figures 2.17 to 2.19 the electrical efficiency of the MGT under partial load is presented for three Capstone MGT models (C600, C800, and C1000). In these three models, at 50% of full load more than 85% of the full load efficiency is achieved. In these three figure, the partial load efficiency of MGTs is compared with typical unrecuperated turbines. It is clear that recuperated MGTs have a considerable advantage over the unrecuperated typical turbine when comparing the partial load operation efficiency.

Recuperator efficiency has a significant impact on the MGT electrical efficiency. For example, a recuperator with 80% effectiveness can increase electrical efficiency

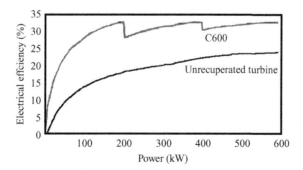

Figure 2.17 Partial load operation impact on the electrical efficiency of a C600 in comparison with an unrecuperated typical turbine [7].

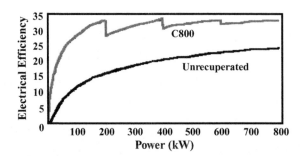

Figure 2.18 Partial load operation impact on the electrical efficiency of a C800 in comparison with an unrecuperated typical turbine [7].

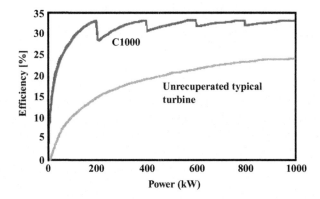

Figure 2.19 Partial load operation impact on the electrical efficiency of a C1000 in comparison with an unrecuperated typical turbine [7].

from 14% to 25%. The impact of recuperator effectiveness on MGT efficiency is presented in Figure 2.20.

More detailed technical data about MGTs are presented in the Tables 2.11 to 2.13. These data include electrical information, fuel system characteristics,

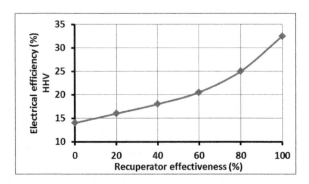

Figure 2.20 MGT efficiency versus recuperator effectiveness [6].

Table 2.11 **Electrical Characteristics of Different MGTs [7]**

Model of MGT	Net Power HP (LP) (kW)	Electrical Efficiency LHV for HP (LP) (%)	Voltage (VAC)	Electrical Service	Frequency (Hz)		Max. Output Current (A)	
					Grid Connection	Standalone	Grid Connection	Standalone
C15	15(-)	23(-)	400–480	3-phase, 4 wire	50/60	10 to 60	23	54
C30	30(28)	26(25)	400–480	3-phase, 4 wire	50/60	10 to 60	46	46
C65	65(-)	28(-)	400–480	3-phase, 4 wire	50/60	10 to 60	100	100
C200	200(-)	33(-)	400–480	3-phase, 4 wire	50/60	10 to 60	290A RMS @400V, 240A RMS @480V	310A RMS
C600	600(-)	33(-)	400–480	3-phase, 4 wire	50/60	10 to 60	870A RMS @400V, 720A RMS @480V	930A RMS
C800	800(-)	33(-)	400–480	3-phase, 4 wire	50/60	10 to 60	1160A RMS @400V, 960A RMS @480V	1240A RMS
C1000	1000(-)	33(-)	400–480	3-phase, 4 wire	50/60	10 to 60	1450A RMS@400V, 1200A RMS @480V	1550A RMS

HP (LP) refers to the high-pressure (low-pressure) fuel system of the MGT.
Nominal full power performance are given at ISO conditions: 59°F, 14.696 psia, 60% RH.
Some utilities may require additional equipment for grid interconnectivity.

Table 2.12 Fuel and Exhaust Characteristics of MGTs [7]

Model of MGT	Fuel/Engine Characteristics				Exhaust Characteristics			
	Natural Gas HHV (MJ/m³)	Inlet Gage Pressure, HP (LP) (kPa)	Fuel Flow HHV, HP (LP) (MJ/hr)	Net Heat Rate LHV, HP (LP) (MJ/kWh)	NO$_x$ at 15% O$_2$ (mg/m³)	NO$_x$ (lb/MWhe)	Exhaust Gas Glow (kg/s)	Exhaust Gas Temperature (°C)
C15	30.7-47.5	379-414(-)	255(-)	15.5(-)	N/A	N/A	N/A	N/A
C30	30.7-47.5	379-414 (1.4-6.9)	457 (444)	13.8 (14.4)	18	0.64	0.31	275
C65*	30.7-47.5	517-552(-)	919(-)	12.9(-)	8	0.17	0.51	311
C200	30.7-47.5	517-552(-)	2400(-)	10.9(-)	18**	0.4**	1.3	280
C600	30.7-47.5	517-552(-)	7200(-)	10.9(-)	18**	0.4**	4	280
C800	30.7-47.5	517-552(-)	9600(-)	10.9(-)	18**	0.4**	5.3	280
C1000	30.7-47.5	517-552(-)	12000(-)	10.9(-)	18**	0.4**	6.7	280

* This C65 has an integrated copper core heat recovery system that produces 120kW hot water with total efficiency LHV of 80%. Heat recovery for water takes place at an inlet temperature of 57 °C (135 °F) and flow rate of 2.5 l/s (40 GPM).
** Low NOx version produces 8 mg/m³ or 0.14 lb/MWhe.
Exhaust emissions for standard natural gas at 39.4 MJ/Nm3 (1,000 BTU/scf) (HHV).
The low-pressure models can use natural gas as fuel while the high-pressure models can use natural gas, landfill gas, digester gas, propane, diesel, and kerosene.

Table 2.13 Dimensions, Weight, Installation Clearances, and Noise of MGTs [7]

Model of MGT	Dimensions W × D × H (m)*	Weight (kg)	Min. Clearance Requirements (m)**				Noise at Full Load at 10 m (db)
			Above	Left & Right	Front	Rear	
C15	0.76 × 1.5 × 1.9	578	0.61	0.76	0.93	0.92	65
C30	0.76 × 1.5 × 1.8	578	0.61	0.76	0.93	0.92	65
C65	0.76 × 2.2 × 2.6	1450	0.61	0.76	1.7	0.76	65
C200	1.7 × 3.8 × 2.5	3413	0.6	1.1	1.1	1.8	65
C600	2.4 × 9.1 × 2.9	15014	0.6	1.5& 0	1.5	2	65
C800	2.4 × 9.1 × 2.9	15558	0.6	1.5& 0	1.5	2	65
C1000	2.4 × 9.1 × 2.9	20956	0.6	1.5& 0	1.5	2	65

* W D H stands for Width × Depth × Height.
** Clearance requirements may increase due to local code considerations.
Height dimensions are to the roofline. The exhaust outlet extends at least 7 in. above the roofline.

exhaust environmental and thermal properties, size dimensions, weight, and noise generation for different MGTs.

As can be seen, MGTs can be classified as high pressure (HP) or low pressure (LP) according to the fuel pressure. The HP-MGTs have higher electrical efficiency and power generation. The fuel inlet pressure in the high-pressure system is from 3.79 bar (g) to 5.52 bar (g) while in the LP-MGT the fuel inlet pressure is only 0.014 to 0.069 bar (g). The exhaust gas temperature ranges from 275 to 311 °C. Minimum required clearances for installation of every MGT unit are also given in the tables. The noise level for all types of MGTs at full load operation is 65 db at a 10 meter distance from the MGT. Electricity frequencies when the MGT is connected to the grid or operating standalone are 50/60 HZ and 10–60 Hz, respectively.

Curve fitting of the data presented in Tables 2.11 to 2.13 results in some useful correlations for simulation and optimization purposes.

$$F(kW) = 3.303E_{nom} + 23.928, 15 \le E_{nom}(kW) \le 1000 \tag{2-18}$$

$$Q_{rec}(kW) = 0.242E_{nom} + 1.545, 15 \le E_{nom}(kW) \le 1000 \tag{2-19}$$

PHR, electrical, and overall efficiencies of MGTs can be derived according to their definitions and the above equations as follows:

$$\eta_e = \frac{E_{nom}}{3.303E_{nom} + 23.928}, 15 \le E_{nom}(kW) \le 1000 \tag{2-20}$$

$$\eta_o = \frac{1.242E_{nom} + 1.544}{3.303E_{nom} + 23.928}, 15 \le E_{nom}(kW) \le 1000 \tag{2-21}$$

$$PHR = \frac{E_{nom}}{0.242E_{nom} + 1.545}, 15 \le E_{nom}(kW) \le 1000 \tag{2-22}$$

The EPA also reported some environmental and techno-economical data about some particular sizes of CHP systems based on MGTs (Tables 2.14 to 2.17).

The emission indices for MGT-CHP units are presented in the Table 2.15 for CO, CO_2, NO_x, and THC.[6] These data are for MGT-CHP working with natural gas and 15% O_2.

The O&M costs depend on the size, fuel type, and technologies used in different components and control systems. In Table 2.16 the full service costs of three engine sizes are presented as an example. A full service includes major overhaul, periodic inspections, and component replacements (such as filters, gages, fuel injectors, etc.). Inspection periods vary with fuel type, technology, and ambient air quality (dust, chemical air pollution, oil vapor, and humidity) at the MGT site. Replacement of components usually is determined by the working hours of the MGT, and component

---

[6]Total hydrocarbon refers to the family of chemicals whose molecules contain carbon and hydrogen atoms.

Table 2.14 Characteristics of some MGT-CHPs [4]

E_{nom} (kWe)	30	65	250
Compressor parasitic power (kW)	2	2	8
MGT cost 2007 $/kW	1290	1280	1410
Total MGT-CHP installed Cost 2007 $/kW	2970	2490	2440
Electric heat rate HHV (Btu/kWh)	15075	13891	13080
$\eta_e(\%)$, HHV	22.6	24.6	26.09
Engine speed (rpm)			
F (MMBtu/hr)	0.422	0.875	3.165
F (kW)	123.6	256.2	926.7
Required fuel gas pressure (psig)	75	75	75
CHP Characteristics			
Exhaust flow (lb/sec)	0.69	1.12	4.7
Exhaust temperature (°C)	277	311	242
Q_{rec} (MMBtu/hr)	0.17	0.41	1.2
Q_{rec} (kW)	50.9	119.5	351.6
Form of recovered heat			
$\eta_o(\%)$, HHV	63.8	71.2	64.0
RFR	0.41	0.47	0.38
PHR	0.55	0.53	0.69
Net heat rate (Btus/kWh)	7313	5796	6882
EEE	46.7	58.9	49.6

Table 2.15 Combustion and Emission Indices for MGT-CHPs [4]

Emissions characteristics			
Electricity capacity (kW)	100	300	800
NO_x (lb/MWh)	0.54	0.22	0.29
CO (lb/MWh)	1.46	0.3	0.14
THC (lb/MWh)	0.19	0.09	0.10
CO_2 (lb/MWh)	1736	1597	1377

Table 2.16 Maintenance Costs of MGT-CHPs [4]

Capacity (kWe)	30	65	250
O&M costs-service contract, 2007 $/kWh	0.015-0.025	0.013-0.022	0.012-0.02

performance. For example, if the pressure loss of the air filters is out of range, they must be cleaned or replaced even if the working hours limit is not satisfied. In addition, if the allowable working hours of a component such as the air filter or fuel filter is reached, it must be replaced even if it operates properly, unless the O&M supervisor gives permission to use those parts until a certain time (for example if the major overhaul time is close, some replacements can be postponed to the overhaul time).

Table 2.17 **Capital Costs of MGT-CHP systems [4]**

Capacity (kW)	30	65	250
All costs in $/kW			
MGT package	1290	1280	1410
Heat recovery and other equipment	430	340	190
Labor/material	710	360	350
Project and construction management	210	200	190
Engineering and fees	210	200	190
Project contingency	90	80	80
Project financing (interest during construction)	30	30	30
Total plant cost	2970	2940	2440

2.2.5 CCHP Based on a Micro-Steam Turbine (MST)

In many processes steam is used, its pressure is reduced by a pressure reduction valve, and then it is condensed. A micro-steam turbine is basically suitable for using the wasted energy of process steam directly or indirectly. Steam also can be produced by a micro-boiler in a Rankine cycle.

The schematics of a CCHP system based on a micro-steam turbine are presented in Figures 2.21 and 2.22. The process steam after generation of power in the micro-steam turbine can be used for cooling or heating purposes as well. An MST can also use the waste heat of process steam indirectly when it is used as an organic Rankine cycle

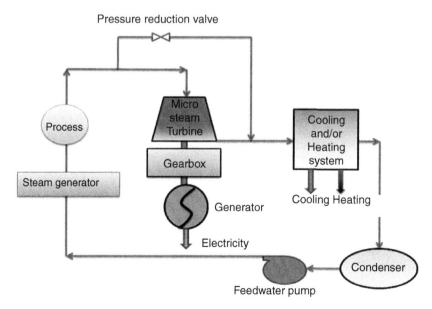

Figure 2.21 A CCHP system based on a micro-steam turbine (direct use of waste heat of process steam).

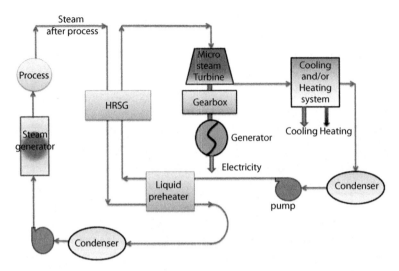

Figure 2.22 A CCHP system based on an ORC micro-steam turbine (indirect use of waste heat of process steam).

(ORC), as is shown in Figure 2.22. In this CCHP cycle the refrigerant is also preheated before entering the heat recovery steam generator (HRSG), in order to as much waste energy as possible.

Energent Corporation has presented a single-stage radial outflow micro-steam turbine with a capacity of 275 kW to convert waste energy to power. The main components of the MST are a radial outflow turbine, epicyclic gearbox, and induction generator. It produces 275 kWe at 50 or 60 Hz AC electricity and its dimensions are W × L × H = 0.86 × 1 × 2 (m). A painting of this MST is presented in Figure 2.23.

The main characteristics of this MST are reported in Table 2.18.

The power output of MST versus the steam flow rate is depicted for three pressure duties (125/15 psig, 125/30 psig, and 200/60 psig) in Figure 2.24.

2.2.6 CCHP Based on Stirling Engine

A Stirling engine is an external combustion reciprocating engine that is able to use various types of fuels. It operates quietly and produces less air pollution. A Stirling engine is classified in three main configurations: α type, β type, and γ type (Figures 2.25 to 2.27). In an α type configuration the engine has two separate pistons and cylinders. The β type engine has a piston and a displacer in one single cylinder, and in the γ type a piston and a displacer reciprocate in separate cylinders. In all three types there are a cooler and heater to contract and expand the working gas, respectively. The working gas may be air, nitrogen, helium, or hydrogen. A CCHP system based on an α type Stirling engine is shown in Figure 2.25. It shows the main components: cylinders, pistons, cooler, heater, crankshaft, gearbox, generator, and cooling and heating systems.

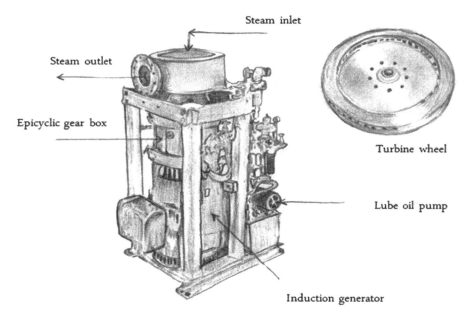

Figure 2.23 Energent MST and its wheel [8]. Painting by Shole Ebrahimi

Table 2.18 **Characteristics of Energent Micro-Steam Turbine [8]**

Efficiency of radial outflow turbine	**80% at pressure ratio 2.5:1**
	70% at pressure ratio 2.5:1
Full load power output	275 kWe@480 V, 60Hz, 3 phase
	275 kWe@400 V, 50Hz, 3 phase
Full load steam output	3690 Btu/kWh
Steam flow rate	4000-20000 lb/hr
Standard working pressure	Maximum: 200 psig
	Minimum: 2 psig
Typical letdown duty	200/60 psig, 150/30 psig, 125/30psig, 125/15 psig
Material/construction	Tolerant of poor-quality steam
	Titanium alloy rotor
	Stainless steel nozzle
Epicyclic gearbox efficiency	97% at 300 kWe shaft
Generator efficiency	96% at 275 kWe
Noise	85 db untreated acoustically
Size and dimensions (W × D × H)	34″ × 42″ × 78″(0.86 × 1.07 × 1.98 m)
PLC control	Single-button startup
	Automatic synchronization
	Hardwired safety trips
Steam generator	Natural gas boiler with 85% efficiency
CO_2 emission productions	1400 lb/MWh
Lifetime	15 years

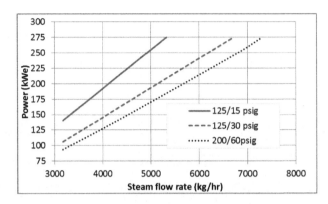

Figure 2.24 Output power of MST versus steam flow rate for three pressure duties [8].

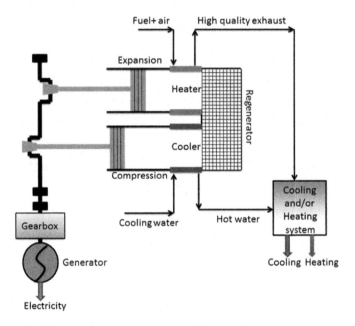

Figure 2.25 A CCHP system based on an α type Stirling engine.

The principals of operation of an α type Stirling engine can be described as follows:

1. The gas is heated by the heater, which results in gas expansion to the maximum volume and pressure reduction (to maintain the maximum constant temperature). This gas expansion pushes the expansion piston back to rotate the crankshaft.
2. As the crankshaft rotates it pushes the expansion piston forward, pushing the gas into the regenerator, leaving its thermal energy in the regenerator to be used later.
3. Then gas is cooled in the cooler, losing its thermal energy and making compression easier. As the expansion piston moves forward, the compression piston moves back to keep the available volume for the gas constant. At this stage the crankshaft mechanism rotates and pushes the compression piston forward to compress the gas and push it back to the regenerator.

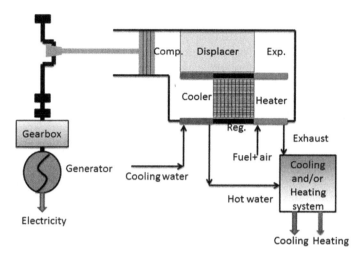

Figure 2.26 A CCHP system based on a β type Stirling engine.

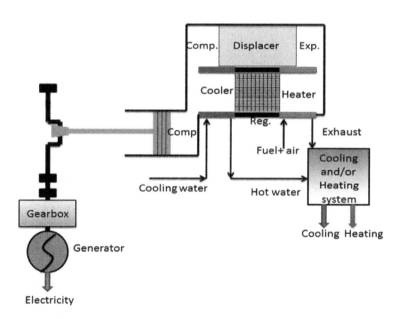

Figure 2.27 A CCHP system based on a γ type Stirling engine.

4. As the gas flows through the regenerator, it captures the thermal energy stored in the regenerator in step 2; its temperature increases and reaches the maximum. The pressure increases as well. This high pressure pushes the expansion piston back, the gas pressure reduces, and this causes the temperature to decrease as well. To avoid more temperature reduction, the gas must be heated to maintain the exit temperature from the regenerator. This completes the cycle and the next cycle starts and ends the same.

As can be seen, the Stirling engine described above can be operated in a closed cycle. Sealing is very crucial for this engine to avoid leakage of working fluid. However, it has no valve such as those that can be found in internal combustion engines. This will result in much less downtime and maintenance costs. In addition, due to external combustion, no combustion product touches the internal parts of the engine, such as the pistons and sealing system. This also helps to create a clean process inside the engine and pressurized gases from the engine will not exhaust into the ambient. This results in less corrosion, oxidation, and energy loss. Heat sources to be recovered for CCHP systems in Stirling engines include heater exhaust gases, hot water leaving the cooler, and lube oil cooler. In the Figures 2.25 to 2.27, the CCHP systems based on an α type, β type, and γ type Stirling engine are presented. In these three CCHP systems, heater exhaust and the hot water of the engine are used to produce the heating and cooling load of the consumer. Since these engines are small, clean, and silent they are very appropriate for residential buildings.

SOLO Stirling 161 [9] is a Stirling engine with a 90°, two-cylinder motor that uses helium as the operating gas and produces maximum electricity of 9.5 kW. Gas is heated to about 650 °C in the heater. The heater includes small tubes that are heated from outside by a burner to a temperature of about 700 °C. Since the Stirling engine is external combustion, many types of fuel can be used. The exhaust gas of the SOLO Stirling engine is about 800 °C. The high-quality exhaust energy is transferred to the combustion air and preheats the combustion air to about 600 °C. The working gas pressure ranges from 40 to 130 bars, leading to adjustment in the output mechanical work from 3 to 9 kW. A magnetic valve is used to adjust the working gas pressure. In Table 2.19 more details of the technical, environmental, and economic data of this engine are presented.

In Figure 2.28, the SOLO Stirling engine is demonstrated.

Figure 2.28 SOLO Stirling 161 [9].

Table 2.19 Characteristics of SOLO Stirling Engine [9]

Dimensions (W × D × H)	1280 × 700 × 980 (mm)
Weight	460 kg
Maximum exit temperature outer circuit	65 °C
Temperature at heating inlet	50 °C
Electrical output capacity	2-9.5 kW
Thermal output capacity	8-26 kW
η_e	22-24.5%
Thermal efficiency	65-75%
η_o	92-96%
Engine type	V2 Stirling engine
Cylinder capacity	160 cm^3
Operating gas	Helium
Max. medium operating pressure	150 bar
Nominal engine speed	1500 rpm
Burner performance, min-max	16-40 kW
Fuel	Natural gas, liquid gas, pellets
Gas line pressure	50 + 15/-5 mbar (g)
Exhaust back pressure, partial/full load	Max. 2 mbar(g)
Exhaust gas temperature	85 °C
Volume of exhaust gas flow	40-100 kg/h
System	Flameless oxidation
Flame control system start/operation	Ionization/temperature
Emission of nitrogen monoxide (NO)	80-120 mg/m^3
Emission of carbon monoxide (CO)	40-60 mg/m^3
Volume of cooling fluid, internal	4.12 liters
Plate heat exchanger (heat recovery)	Stainless steel, copper soldered
Cooling water flow via external pump	0.5-2 m^3/hr
Cooling water pressure	3 bar
Voltage	400 V
Frequency	50 Hz
Phases	3
Stating current	25 A
Operating current	15.5 A
Fuel consumption (net calorific value)	1.2-3.8 Nm^3/hr
NOx emission(at 5% O_2)	80-120 mg/m^3
CO emission (at 5% O_2)	50 mg/m^3
Capital investment cost of unit	About 25000 €
Specific cost of unit	12500-2632 €/kWe
Service intervals	4000-6000 hrs
Average hours of operation (full load-partial load)	5800-7800 hrs

The electrical efficiency of the SOLO Stirling engine increases with helium pressure, but the overall efficiency remains the same. Both the heat and electrical outputs increase as the helium pressure increases from 6 MPa to 12 MPa. The characteristic curves are illustrated in Figure 2.29.

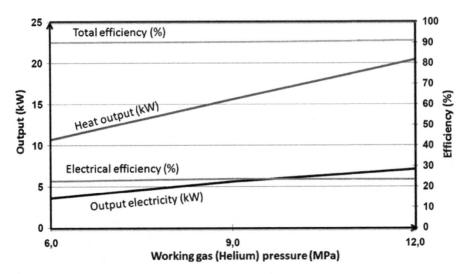

Figure 2.29 Electricity, heat, and output efficiency of SOLO Stirling 161 versus helium pressure (SOLO data measured 2001-09-24) [10].

Table 2.20 DTE Energy Stirling Engine Characteristics [6]

Capacity (kWe)	20	25
$\eta_e(\%)$	29.6	29.6
$\eta_o(\%)$	82	82
NO_x (gm/bhph)	o.29 (standard), 0.15 (Ultra low emission)	o.29 (standard), 0.15 (Ultra low emission)
CO (gm/bhph)	o.32(standard and ultra low emission)	o.32(standard and ultra low emission)

DTE Energy is also a main producer of Stirling engines with capacity of up to 1 MW, and overall efficiency of 84%. The important characteristics of the 20 kWe and 25 kWe Stirling units of DTE Energy are tabulated in Table 2.20.

Some of the other companies that are working in the Stirling engines field are listed below [6]:

- WhisperTech has developed a 1 kWe Stirling engine, with η_e of 12%.
- Sigma developed a 3 kWe Stirling engine with 9 kW thermal output and η_e higher than 25%.
- Sunpower Inc. developed a 7 kWe Stirling engine working with biomasses such as wood, wood pellets, sawdust, and chips.
- Both Enatec and BG Group developed a 1 kWe free piston Stirling engine with electrical efficiency of 16%.

2.2.7 CCHP Based on Fuel Cells

Fuel cells are compact, clean, silent, environmentally friendly, and energy efficient, but not yet cost effective. Fuel cells do not produce power by combustion of fuel and generating reciprocating or rotational motions in a crankshaft or rotor. They have no moving parts. They produce direct current (DC) electricity and heat through thermochemical reaction of a fuel and passing charged ions from an anode electrode to the cathode electrode through an electrolyte. To create a permanent current of electricity, fuel must always flow to the anode side. On the cathode side the oxidizer, which is oxygen (taken from air flow), must always flow adjacent to the cathode.

The reaction that occurs on the anode is as follows:

$$2H_2 \rightarrow 4H^+ + 4e^- \quad (Anode\ reaction) \tag{2-23}$$

The electrons flow through a wire and produce electricity as the first product, while the hydrogen ions flow through the electrolyte and meet the free electrons from the wire and oxidizer on the cathode to complete the cathode reaction as follows:

$$O_2 + 4H^+ + 4e^- \rightarrow 2H_2O \quad (Cathode\ reaction) \tag{2-24}$$

These two reactions create the overall reaction, which takes place in the fuel cell, producing heat and water vapor as the second and third products of a fuel cell:

$$O_2 + 2H_2 \rightarrow 2H_2O + heat \quad (Overall\ fuel\ cell\ reaction) \tag{2-25}$$

The amount of heat released in this reaction equals the Gibbs free energy of the products minus the Gibbs free energy of the reactants:

$$heat = \sum_{i=1}^{m}(N_i g_i)_{products} - \sum_{j=1}^{n}(N_j g_j)_{reactants} \tag{2-26}$$

where N is the mole number of each component and g is the Gibbs free energy of the component.

In the steam reforming reaction the vaporized hydrocarbon fuel reacts with the superheated steam and creates H_2 and CO. Carbon monoxide takes part in the water-gas shift reaction and results in the production of hydrogen and carbon dioxide. As a result the H_2 concentration has increased and CO production has decreased. The steam reforming reaction and water-gas shift reaction are as follows. The fuel is assumed to be CH_4.

$$\begin{aligned}
CH_4 + H_2O + heat &\leftrightarrow CO + 3H_2 \quad (Steam\ reforming\ reaction) \\
CO + H_2O &\leftrightarrow H_2 + CO_2 + heat \quad (Water - gas\ shift\ reaction)
\end{aligned} \tag{2-27}$$

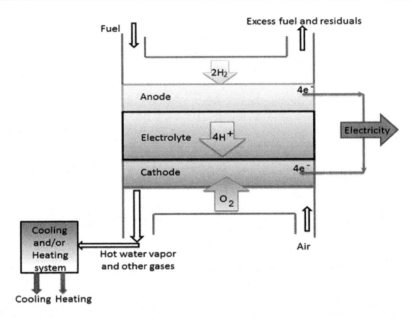

Figure 2.30 A CCHP system based on a fuel cell.

The total heat can be used in thermally activated cooling systems or heating systems to produce the cooling or heating loads of the target consumer. Figure 2.30 shows the application of a fuel cell in CCHP technology.

There are five types of fuel cells: proton exchange membrane (PEMFC), phosphoric acid (PAFC), alkaline (AFC), molten carbonate (MCFC), and solid oxide (SOFC). Each fuel cell has a different charge carrier and electrolyte. A catalyst, which may be platinum, nickel, or parasites, usually improves the reaction. Different types of fuels such as hydrocarbons, methanol, hydrogen, hydrazine, alcohols, natural gas, propane, and diesel can be used. Their size ranges from 3 kW to 10 MW; therefore they can be used for different purposes from a single-family home to large complexes, hospitals, universities, etc. The operational temperature in fuel cells is very different and ranges from 50 to 1000 °C. The electrical efficiency also varies from about 30% to 70% depending on the fuel cell type. Their cost is still high in comparison with other prime movers. Different types of fuel cells are compared from technical, environmental, and economic points of view in Tables 2.21 to 2.23.

Fuel cells are very flexible in partial load operation. They maintain a high efficiency even when the load decreases to about one-third of the nominal load. This is due to the stack behavior, since its operation improves in partial loads. Figure 2.31 shows the partial load efficiency of a PAFC with a nominal size of 200 kWe.

Fuel cells like other prime movers are rated at ISO conditions of 1 bar and 25 °C. The efficiency and output power of fuel cells decrease as the ambient temperature and altitude increases. This is mainly due to the decreasing efficiency of ancillary equipment such as blowers and compressors [4].

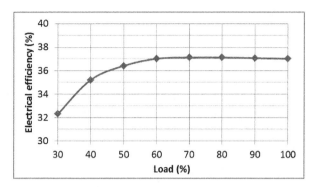

Figure 2.31 Partial load operation of a 200 kWe PAFC [4].

Table 2.21 Comparison of Different Types of Fuel Cells from a Technical Point of View [4, 6]

Fuel Cell Type →	PEMFC	AFC	PAFC	MCFC	SOFC
Charge carrier	H^+ ions	OH^- ions	H^+ ions	$CO_3 =$ ions	$O^=$ ions
Electrolyte type	Polymeric membrane	Aqueous potassium, KOH	Phosphoric acid solutions	Phosphoric acid (immobilized liquid), molten alkali carbonates	Stabilized zirconia ceramic matrix with free oxide ions
Electrode material	Porous carbon	Metals	Porous carbon or graphite	Nickel or nickel oxide	Ceramic
Typical construction	Plastic, metal, or carbon	Plastic, metal	Carbon porous ceramic	High-temperature metal, porous ceramic	High-temperature metal, ceramic
Catalyst	Platinum	Platinum, nickel, metal oxides, or noble metals	Platinum	Nickel or nickel oxide	Parasites, ceramic
Interconnection material	Carbone or metal	Metal	Graphite	Nickel or stainless steel	Nickel, ceramic, or stainless steel
Prime cell components	Carbone	Carbone	Graphite	Stainless steel	Ceramic
Oxidant	Air or O_2	Purified air or O_2	Air or O_2	Air	Air

(Continued)

Table 2.21 **Comparison of Different Types of Fuel Cells from a Technical Point of View [4, 6]** *(cont.)*

Fuel Cell Type →	PEMFC	AFC	PAFC	MCFC	SOFC
Fuel	Hydrocarbons or methanol	Clean hydrogen or hydrazine	Hydrocarbons or alcohols	Clean hydrogen, natural gas, propane, or diesel	Natural gas or propane
Operational temperature	50-100 °C	50-220 °C	100-200 °C	600-700 °C	500-1000 °C
Size range	3-250 kW	10-200 kW	100-200 kW	250 kW-5 MW	1-10 MW
Electrical efficiency*	30%-50%	32%-70%	40%-55%	55%-57%	50%-60%
Primary contaminants	CO, sulfur, and NH_3	CO, CO_2, H_2S, CH_4,	CO > 1%, H_2S	H_2S	H_2S
Internal fuel reforming	No	No	No	Yes	Yes
Applications	Vehicles, mobile, CHPs and CCHPs	Space vehicles	Medium-scale stationary, CHPs and CCHPs	Large- and average-size power stations, hybrid,** CHPs and CCHP	All sizes of power stations, CHPs and CCHPs

* The efficiencies are based on values for hydrogen fuel, and do not include the electricity for hydrogen reforming.
** Hybrid fuel cell-turbine combined cycle systems to achieve system electrical efficiencies in excess of 70 percent HHV [4].

Table 2.22 **Comparison of Different Types of Fuel Cells from an Environmental Point of View [6]**

Fuel Cell Type →	PEMFC	PEMFC	PAFC	MCFC	SOFC
E_{nom} (kWe)	10	200	200	250	100
η_e (%) HHV	30	35	36	46	45
NO_x (ppmv at 15% O_2)	1.8	1.8	1.0	2.0	2.0
NO_x (lb/MWh)	0.06	0.06	0.03	0.06	0.05
CO (ppmv at 15% O_2)	2.8	2.8	2.0	2.0	2.0
CO (lb/MWh)	0.07	0.07	0.05	0.04	0.04
CO_2 (lb/MWh)	1360	1170	1135	950	910
Carbon (lb/MWh)	370	315	310	260	245
Unburnt hydrocarbons (ppmv at 15% O_2)	0.4	0.4	0.7	0.5	1.0
Unburnt hydrocarbons (lb/MWh)	0.01	0.01	0.01	0.01	0.01

Table 2.23 **Comparison of Different Types of Fuel cells from an Economic Point of View [4, 6]**

Fuel Cell Type →	PEMFC	PEMFC	PAFC	MCFC	SOFC
E_{nom} (kWe)	10	200	200	250	100
Total installed investment cost (2002 $/kWe)	5500	3600	4500	5000	3500
Total O&M cost ($/kWh)	0.033	0.023	0.029	0.043	0.023

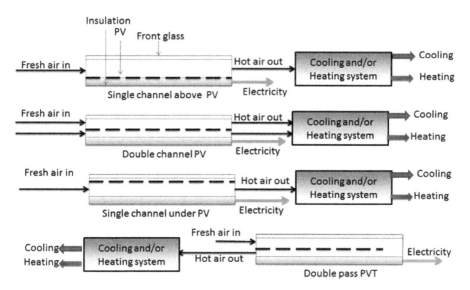

Figure 2.32 Schematics of some PVT-CCHP configurations.

2.2.8   CCHP Based on Photovoltaic Thermal (PVT)

PVT is a CHP technology that integrates solar heating and solar electricity. This technology reduces the needed installation area, and increases the efficiency of using solar
energy. A schematic of this technology is presented in Figure 2.32.

A PVT generates 200% to 300% more energy per unit of area as compared to PV
alone. PVT has a total operating efficiency of above 50%, which is significantly higher
than the PV standalone with an efficiency of about 8% to 15%. A PV standalone
system has a very long payback period due to its low efficiency, but PVT systems
decrease fuel consumption and as a result considerable money savings occurs and the
payback period decreases as well. The air passing though the PVT can reach temperatures as high as 90 °C, which is suitable for running a single-effect absorption chiller,
adsorption chiller, and desiccant cooling system as well. The SolarWall Company[7]
produces PVT systems that generate 300-400 W/m^2 while PV standalone electrical
output is about 100 W/m^2.

2.3   Thermally Activated Cooling Systems

Thermally activated cooling (TAC) systems, which are also called thermally driven
cooling systems or heat powered cooling systems, are the most appropriate technologies for integration with CHP systems to build CCHP systems. These systems use heat
instead of electricity to produce cooling. The required heat can be fully or partially

[7]Source: www.SolarWall.com.

provided by recovering the waste heat of prime movers. Among these systems single-, double-, and triple-effect absorption chillers (H_2O-LiBr and NH_3-H_2O types), adsorption chillers (silica gel-H_2O, activated carbon-methanol, and $CaCl_2$-NH_3 types) and desiccant cooling systems (liquid and solid types) are available for utilization in CCHP applications. TAC systems need heat sources with different temperatures that can be provided by the waste heat, and, if needed, an auxiliary heater can be used to compensate for the lack of the required input energy. An ejector refrigeration cycle is another option that can be used in conjunction with some power cycles to produce cooling demand.

The thermal coefficient of performance (COP_{th}) of TAC systems can be calculated as the output refrigeration (C_{out}) divided by the input thermal energy (H_{in}):

$$COP_{th} = \frac{C_{out}}{H_{in}} \qquad\qquad (2\text{-}28)$$

The COP of the TAC considers the electricity consumption (E_{in}) of the TAC system as well, and is formulated as follows:

$$COP = \frac{C_{out}}{H_{in} + E_{in}} \qquad\qquad (2\text{-}29)$$

In the following, we will discuss the principals of operation and the important characteristics of TAC systems sequentially.

2.3.1 Absorption Chillers: H_2O-LiBr

Absorption is the process in which a substance with a particular phase is integrated with another substance with a different phase to form a solution. For example, in LiBr-H_2O absorption chillers, water vapor is absorbed by the lithium bromide liquid in the absorber.

LiBr-H_2O absorption chillers use H_2O as the refrigerant and lithium bromide as the absorbent. Since absorption chillers use thermal compression, they have less moving parts (and consequently require less maintenance) with respect to the mechanical compression equipment such as different screw, scroll, reciprocating, or centrifugal compressors. The electricity consumption of absorption chillers is negligible (some liquid pumps require electricity), hence they reduce electricity peak demand in summer, especially in hot climates. They are also friendly to the environment because they do not use atmosphere-harming refrigerants. Moreover, since journal (Babbitt) bearings and sealing systems of pumps use the working fluid for lubrication and cooling, no lubricant is used and degradation of the refrigerant and cooling capacity reduction do not occur. On the contrary, in some of the mechanical compression equipment, lubricants are in direct contact with the refrigerant, which may result in contamination of the refrigerant and reduction of the cooling capacity of the cooling system.

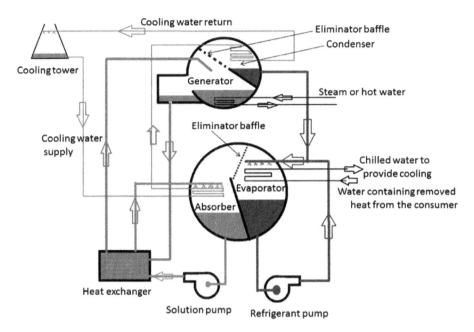

Figure 2.33 Schematic of the basic components of a single-effect H₂O-LiBr absorption chiller.

The main components of absorption chillers and their functions are discussed sequentially (see Figure 2.33)

Generator: The generator includes bundles of tubes through which hot water, steam, or combustion products flow, evaporating and desorbing the refrigerant (water vapor) from the dilute absorbent solution (rich refrigerant solution).

For generators that use hot water or steam as the heating source, the associated chillers are called indirect fired absorption chillers. If the generator uses a burner and combustion for heating purposes, the generator is called direct fired. In addition, if refrigeration desorption occurs in one, two, or three stages (generators) the associated chiller is called single, double, or triple effect, respectively. Single-effect chillers mostly use hot water (70-90 °C) or steam (2-3 bar) or the exhaust of prime movers (250-300 °C). Double-effect chillers use hot water (120-170 °C), steam (4-8 bar), the exhaust of prime movers (400-500 °C), or are direct fired as well. Triple-effect chillers are mostly direct fired or use hot water (200-230 °C). The pressure inside the generator due to evaporation of the refrigerant reaches about 6 kPa in single-effect chillers. In double-effect chillers the vapor pressure in the first and second stages of the generator is about 130 kPa and 8 kPa, respectively [11, 12].

Separator-eliminator: The generator and condenser are usually packed inside a cylindrical vessel, with the generator beneath the condenser and some plates and eliminators separating the generator from the condenser. The desorbed refrigerant in the generator passes through the droplet eliminator to enter the condenser. The eliminator prevents the solution droplets, which may contain absorbent, from entering the condenser. This helps to create pure liquid refrigerant in the condenser.

Condenser: The refrigerant vapor, after passing through the eliminators, flows over a bundle of tubes of the condenser that contain cooling water. The cooling water, which is returning from the cooling tower with a temperature of about 30-32 °C, absorbs the heat of the refrigerant vapor and condenses it for preparation for use in the evaporator for cooling purposes in the next step [12]. The cooling water leaves the chiller at a temperature of about 37 °C and enters the cooling tower to lose heat by sensible and latent mechanisms to reach the appropriate temperature (30-32 °C) for use in the cooling water chiller circuit

Evaporator: The condensed refrigerant in the condenser, after passing through a liquid trap or orifice, flows to the distributers to spray the refrigerant over the outside surface of water tubes. These water tubes contain the water carrying the cooling demand (the heat should be removed) coming back from the consumer. The water is supposed to be chilled to a temperature of 5-10 °C to provide the cooling demand of the consumer [11]. Since the inside of the evaporator is vacuumed (0.8-20 kPa) the liquid refrigerant starts boiling as it contacts with the water tubes from the consumer. A portion of the refrigerant is vaporized and the remaining liquid refrigerant is stored at the bottom of the evaporator. The vaporized refrigerant goes to the absorber (just beside the evaporator in the same vessel) to be absorbed by the diluted refrigerant solution. The liquid refrigerant at the bottom of the evaporator is pumped back to the distributer to be sprayed on the water tubes inside the evaporator. The water tube loses heat and the water chills.

Absorber: The absorber, which is separated from the evaporator by a longitudinal baffle, is responsible for absorbing the vaporized refrigerant coming from the evaporator using the hot poor refrigerant solution coming from the generator. The hot poor refrigerant solution passes through a heat exchanger before entering the absorber, and loses some of its heat. A cooling water line is also used to condense the refrigerant vapor and accelerate the absorbing process. This final liquid solution, which is rich refrigerant, is pumped from the bottom of the absorber to enter the generator. But before entering the generator it passes through a heat exchanger for preheating by the hot poor refrigerant solution coming from the generator to the absorber. When the preheated rich refrigerant solution enters the generator the cycle completes. Figure 2.33 shows the procedure described above.

Reference [12] proposes general characteristics for single- and double-effect H₂O-LiBr absorption chillers (Tables 2.24 and 2.25).

The main producers of absorption chillers are Carrier, YORK, TRANE, Robur, McQuay, Rotartica, LG Machinery, Century, Broad, Entropie, Colibri, Sanyo, Mitsubishi, Ebara, Yazaki, Shuangliang Eco-Energy, Kawasaki, Thermax, Kyung Won Century, and Hitachi.

Hitachi Appliances, Inc. has produced single-effect high-temperature (higher than 140 °C) and low-temperature (80-90 °C) hot water absorption chillers. This company has also produced gas/oil fired absorption chiller-heater with capacities from 422 kW to 4922 kW and COPs of 1.12-1.43 (gas fired) and 1.12-1.49 (oil fired). Steam double-effect absorption chillers from Hitachi consume 3.5-4.5 kg/hr per refrigeration ton. Their capacity ranges from 422 to 19690 kW [14].

Table 2.24 Characteristics of a Single-Effect H₂O-LiBr Absorption Chiller [12]

Performance Characteristics	
Steam input pressure	60 to 80 kPa (gage)
Steam consumption (per kilowatt of refrigeration)	**1.48 to 1.51 kW**
Steam input temperature	**115 to 132 °C; as low as 88 °C for some smaller machines for waste heat applications**
Heat input rate (per kilowatt of refrigeration)	**1.51 to 1.54 kW; as low as 1.43 kW for some smaller machines**
Cooling water temperature in	**30 °C**
Cooling water flow (per kilowatt of refrigeration)	**65 mL/s; up to 115 mL/s for some smaller machines**
Chilled water temperature	**6.7 °C**
COP	**0.7-0.8**
Chilled water flow (per kilowatt of refrigeration)	**43 mL/s; 47 mL/s for some smaller international machines**
Electric power (per kilowatt of refrigeration)	**3 to 11 W with a minimum of 1 W for some smaller machines**
Physical Characteristics	
Nominal capacities	**180 to 5800 k; 18 to 35 kW for some smaller machines**
Length	**3.3 to 10 m; as low as 0.9 m for some smaller machines**
Width	**1.5 to 3.0 m; 0.9 m minimum for some smaller machines**
Height	**2.1 to 4.3 m;1.8 m for some smaller machines**
Operating mass	**5 to 50 Mg; 320 kg for some smaller machines**

Table 2.25 Characteristics of a Double-Effect H₂O-LiBr Absorption Chiller [12,13]

Performance Characteristics	
Steam input pressure	**790 kPa (gage)**
Steam consumption (condensate saturated conditions) (per kilowatt of refrigeration)	**780 to 810 W**
Steam input temperature	**188 °C**
Heat input rate (per kilowatt of refrigeration)	**0.83 kW**
Cooling water temperature in	**30 °C**
Cooling water flow (per kilowatt of refrigeration)	**65 mL/s to 80 mL/s**
Chilled water temperature	**7 °C**
COP	**1.1-1.45**
Chilled water flow (per kilowatt of refrigeration)	**43 mL/s**
Electric power (per kilowatt of refrigeration)	**3 to 11 W**
Physical Characteristics	
Nominal capacities	**350 to 6000 kW**
Length	**3.1 to 9.4 m**
Width	**1.8 to 3.7 m**
Height	**2.4 to 4.3 m**
Operating mass	**7 to 60 Mg**

YORK (since the 1950s) produces single-effect water-LiBr absorption chillers with a cooling capacity ranging from 422 kW to 4840 kW. Hot water chillers from this company operate with an entering hot water temperature of 80 to 128 °C. Steam-operated chillers run with an inlet steam pressure of 0.2 bar (g) to 0.95 bar (g). The COP of units reaches 0.72, and the fuzzy-logic control system is able to maintain the leaving chilled water temperature set point at 10% to 100% of load. Service inspections are recommended for every 50,000 hours of operation (see Figure 2.34) [15].

Partial load operation of absorption chillers results in more energy consumption and COP reduction. For example Figure 2.35 shows the partial load energy consumption of a typical absorption chiller. As the figure shows, at a partial load of 20%, the consumption reaches about 145% of full load consumption for production of the same amount (20% of full load) of cooling. The energy consumption for a partial load of 20% has increased about 9% ($0.2 \times (1.45 - 1)$) and the COP has decreased about 8.3% (COP/1.09).

Rotartica has produced a solar single-effect water-LiBr absorption chiller with a capacity of 4.5 kW. The inlet hot water temperature to the chiller is 90 °C with a

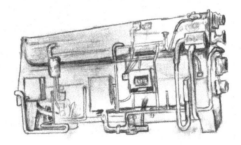

Figure 2.34 Single-effect absorption chiller [15]. Painting by Shole Ebrahimi

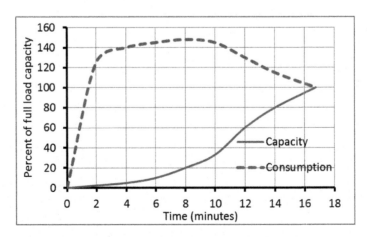

Figure 2.35 Partial load energy consumption for a typical single-effect absorption chiller [16].

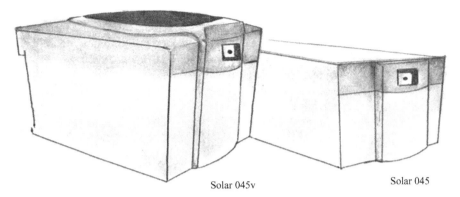

Figure 2.36 Rotartica solar absorption chiller models [17]. Painting by Shole Ebrahimi

flow rate of 15 liters/min. The chilled water temperature is 12 °C with a flow rate of 12 liters/min, and the returning warm water temperature is 35 °C with a flow rate of 33 liters/min. The COP of these chillers approaches 0.8 at certain conditions, and their lifetime is estimated at 15 years. Paintings of two Rotartica chiller models, the Solar 045 and Solar 045v are presented in Figure 2.36 [17].

2.3.2 Absorption Chillers: NH₃-H₂O

Ammonia-water absorption chillers use ammonia as the refrigerant and water as the absorber. The main components and flow diagram of a typical NH_3-H_2O absorption chiller is demonstrated in Figure 2.37. According to the cycle, the solution inside the

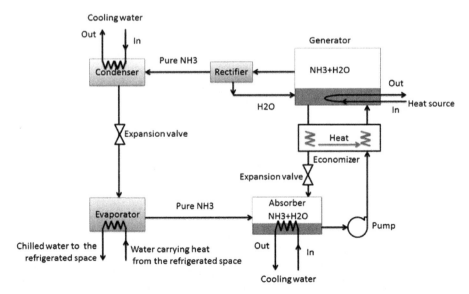

Figure 2.37 Schematic of a single-effect NH_3-H_2O absorption chiller.

generator is heated and some of the solution is vaporized. The vapor, which is rich in ammonia, enters the rectifier; the rectifier separates water and returns it to the generator. The pure ammonia enters the condenser to be liquefied by the cooling water (or air cooler) in order to increase its ability to receive heat before vaporizing in the evaporator. The liquid ammonia then passes through an expansion valve to lose its pressure while maintaining its enthalpy. As the liquid ammonia with low pressure enters the evaporator, it contacts with the outer surface of water tubes, and starts boiling as a result of taking heat from the returning water line from the consumer (the returning water line has collected heat from the space, which should be refrigerated). The water inside the tubes loses heat and chills. The vapor ammonia leaves the evaporator and enters the absorber while it is being cooled by the cooling water. Ammonia dissolves and reacts with water to form NH_3+H_2O in an exothermic reaction while some heat is released [18].

The amount of NH_3 that dissolves in water depends on the temperature of the absorber. The lower the temperature of absorber is the more ammonia will be dissolved. Therefore, cooling water constantly cools the absorber. The liquid water-ammonia solution, which is rich in ammonia, is pumped back to the generator and preheated in the economizer before entering the generator. The cycle is now complete.

NH_3-H_2O absorption chillers exist in single-effect, double-effect, and GAX types. The GAX stands for generator/absorber heat exchanger. In a GAX cycle the absorber heat is used to heat rich ammonia solution, which is pumped to the generator and the low-temperature section of the generator [13]. The dependency of the evaporation temperature of NH_3, heat source temperature, and cooling water temperature is illustrated in Figure 2.38. According to this picture, by decreasing the cooling water temperature and increasing the heat source temperature, an evaporation temperature of as low as -50 °C is achievable.

Table 2.26 summarizes NH_3-H_2O absorption chiller characteristics [3, 11, 12].

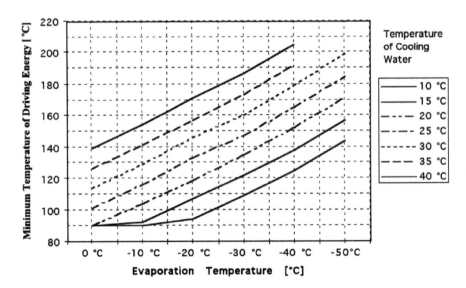

Figure 2.38 Minimum temperature of heat source for different cooling water temperatures [11].

Table 2.26 Characteristics of NH₃-H₂O Absorption Chillers [3, 11, 12]

Characteristics	Single-Effect	Single-Effect	GAX	Double-Effect
Cooling temp. **Heat source temp.**	-60 to 0 (°C) 100-200 (°C)	5-10 (°C) 80-120 (°C)	<0 (°C) 160-200 (°C)	<0 (°C) 170-220 (°C)
COP_{th}	0.25-0.6 (GAX more than 0.7)	0.5-0.6	0.7-0.9	0.8-1.2
Capacity (kW)	10-6500	10-30	10-90	Up to 3500
Manufacturers	Colibri-Stork	Apina, SolarNext of Germany, Colibri-Stork, Hans Güntner GmbH	Cooling Technologies Inc., Colibri-Stork, Robur Company	
Current status	Commercial	Commercial	Small-batch commercial	Experimental
Application **Remarks**	Industrial Rectification of refrigerant is required. Working solution is environmentally friendly. Operating pressure as high as with NH₃. No crystallization problem. Suitable for use as heat pump due to wide operating range.	Residential Can be used for residential cooling applications. Has the potential of being air cooled. Innovative solution pumps and heat exchangers. Rectification of refrigerant is required. Working solution is environmentally friendly. Operating pressure as high as with NH₃. No crystallization problem. Suitable for use as heat pump due to wide operating range.	Industrial Direct fired or requires a higher temperature heat source. COP_{th} is better than single effect. Has the potential of being air cooled. Rectification of refrigerant is required. Working solution is environmentally friendly. Operating pressure as high as with NH₃. No crystallization problem. Suitable for use as heat pump due to wide operating range.	Industrial Safety problem due to high ammonia pressure. Heat release from the first stage absorber used for the second stage generator. Rectification of refrigerant is required. Working solution is environmentally friendly. Operating pressure is high as using NH₃. No crystallization problem. Suitable for using as heat pump due to wide operating range.

Absorption chiller installation costs vary from 140 to 290 \$/kW and yearly O&M cost is also 4.5 to 9 \$/kW [3]. The following equation is presented in [19] for the capital cost of absorption chillers:

$$I_{abc} = \begin{cases} -81.552\ln(C_{nom}) + 778, & C_{nom} > 1000 \ kW \\ -35.4\ln(C_{nom}) + 431, & C_{nom} < 1000 \ kW \end{cases} \tag{2-30}$$

2.3.3   Adsorption Chillers

Adsorption technology is new and environmentally friendly. These systems can use low quality waste-heat for cooling production in different residential and commercial buildings. They can also be used in conjunction with absorption chillers if there are heat sources with different qualities.

Adsorption chillers, like the absorption ones, use working pair materials, which in this case are called adsorbent/adsorbate. The adsorbent is a solid bed that takes in and releases the adsorbate vapor in different stages. The most common working pair in adsorption chillers is silica gel–water. Silica gel is the adsorbent and water is the adsorbate. Adsorption and desorption are two important phenomena that occur due to cooling and heating of the adsorbent bed, respectively. In adsorption, the adsorbent takes in the adsorbate vapor, and in desorption the adsorbent releases the adsorbate vapor due to heating. In this stage the adsorbent is regenerated for a new cycle of adsorption. The main components and flow diagram of an adsorption chiller are demonstrated in Figure 2.39.

According to Figure 2.39A, as the liquid adsorbate with low pressure and temperature sprays over the outer surface of water tubes returning from the consumer, it starts boiling and evaporating by taking heat from the returning water tubes. As a result the water is chilled and pressure inside the evaporator increases. At this stage, since the adsorption vessel pressure is less than that of the desorption vessel (due to releasing vapor in the desorption vessel), the absorption vessel valve (V1) opens automatically and adsorbate vapor enters the adsorption vessel. This flow of vapor continues until the adsorbent bed is saturated with vapor and cannot take in any more vapor. As the pressure inside the adsorption vessel increases, it closes V1 and at the same time opens V2 and closes V3 (Figure 2.39B). Now the pump reverses the flow and switches between the adsorption and desorption vessels. This time since the desorption vessel is releasing vapor (Figure 2.39B) the pressure inside this vessel increases and V1 remains closed while V2 is opened. In the evaporator, due to adsorbate evaporation, the pressure increases again and opens V4 to send vapor to the adsorption vessel. During the adsorption of vapor, V3 remains closed until the adsorbent bed is saturated with vapor again. The pressure inside the vessel increases and the cycle will be repeated again.

Different working pairs are used in various adsorption technologies. Table 2.27 presents the characteristics of different working pairs used in adsorption chillers. The most common type of adsorption chillers work with the silica gel–water pair; this

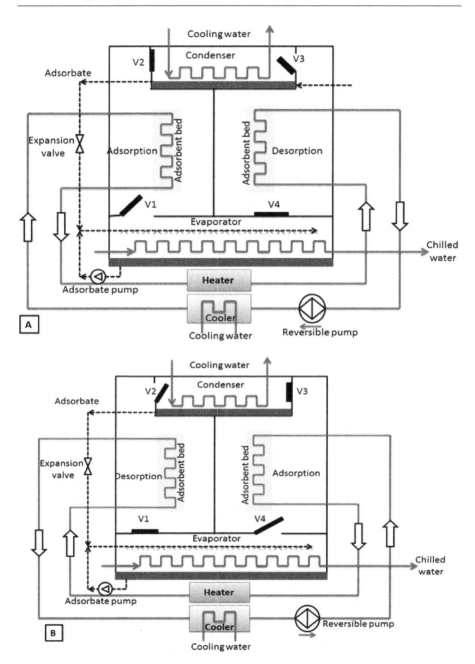

Figure 2.39 Adsorption chiller components and principals of operation.

Table 2.27 **Characteristics of Working Pairs Used in Adsorption Chillers [3, 11, 13]**

Adsorbent	Adsorbate (Refrigerant)	COP	Heat of Adsorption (kJ/kg)	Toxicity	Vacuum Level	Release Temp. (°C)	Heat Sources	Applications
Silica gel	H_2O	0.56-0.6	2800	No	High	60-90	Solar energy, low-temperature recoverable waste	Space cooling, refrigeration
	CH_3OH		1000-1500	Yes	High			
Zeolite	H_2O	0.55-0.69	3300-4200	No	High	>150	High-temperature waste heat	Space cooling, refrigeration
	NH_3		4000-6000	Yes	Low			
Activated charcoal	C_2H_5OH		1200-1400	No	Moderate	80	Solar energy, low-temperature waste heat	Low-temperature ice making
$CaCl_2$	CH_3OH		1800-2000	Yes	High	80	Solar energy, low-temperature waste heat	
	NH_3		1386	Yes	Low	95-120		Low-temperature ice making
Charcoal fiber	CH_3OH		N/A	Yes	Low			
			>2000	Yes	High	120		

technology works with a heat source temperature below 100 °C, which makes this technology very appropriate for CCHP systems. Since it is not toxic it can also be used in residential applications as well [3, 11].

The first producer of adsorption chillers, Nishiyodo Kuchouki, Co. Ltd, produced adsorption units with cooling capacities from 51 to 1000 kW, working with hot water temperatures from 50 to 90 °C and chilled water temperatures approaching 3 °C. A thermal COP of 0.7 is also achievable when working with 90 °C hot water. Mayekawa is another producer of silica gel–water adsorption units. The adsorption chiller of this company works with 75 °C hot water, provides 14 °C chilled water, and has a COP of 0.6. This company has also produced zeolite-water adsorption chillers with capacities of 105, 215, and 430 kW (0.0043 to 0.0067 kWe per kW of cooling capacity). The heat source (hot water) temperature is 68 °C, the cooling water is 27 °C, and the chilled water temperature is 15 °C. The flow rates of hot water for the three sizes are 20, 40, and 80 m³/hr, respectively. In addition the flow rates of chilled water for the three sizes are 12.1, 24.3, and 48.7 m³/hr, respectively. The overall power consumed by the refrigerant and vacuum pumps for the three sizes is 0.7, 0.95, and 1.85 kW, respectively. The sizes of the three units (L × W × H) are 3.7 × 1.5 × 2.8 (m), 3.7 × 2.45 × 2.8 (m), and 6.1 × 2.45 × 2.8 (m), respectively.

Figure 2.40 shows a zeolite adsorption chiller from Mayekawa [21].

Mycom also has produced silica gel–water adsorption chillers with cooling capacities from 50 to 350 kW. This chiller has a thermal COP of 0.6 and provides 9 °C chilled water while working with 75 °C hot water. SorTech is another producer that has produced a 5.5 kW adsorption chiller [3, 11].

With regard to the cost of adsorption chillers, [3] reports that for a series of these units with capacities of 10, 20, 50, and 100 kW capital costs could be 10,000, 15,000, 30,000, and 50,000 US dollars, respectively.

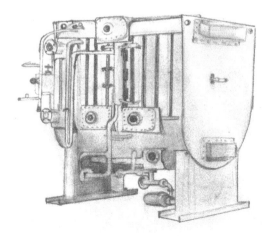

Figure 2.40 The zeolite-water adsorption chiller from Mayekawa [21]. Painting by Shole Ebrahimi

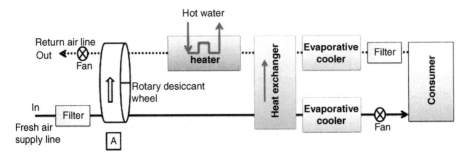

Figure 2.41 Solid desiccant cooler cycle.

2.3.4 Desiccant Cooling

Desiccant cooling is also another option that can be used in CCHP systems. Desiccant dehumidifiers, which are the heart of these cooling systems, need to be regenerated periodically to continue working. According to the temperature required for regeneration of the desiccant, the waste heat of different prime movers can be utilized in these systems. Desiccant cooling may be liquid type (absorption systems) or solid type (desorption systems). In Figure 2.41, a solid desiccant cooling system that uses a rotary wheel for the desiccant is presented. In the supply line, fresh air is pulled in by a fan. Since the desiccant has been regenerated, it is still hot; therefore as air moves through the rotary desiccant wheel, it leaves its humidity in the desiccant and at the same time captures heat from it. Dry and warm air then passes through a sensible heat exchanger to preheat the retuning air, which is supposed to be used for regeneration of the desiccant. As the supply air loses some of its energy in the heat exchanger, it enters the evaporative cooler to be further cooled and humidified. On the return line from the consumer, air first enters the evaporative cooler to be cooled and increase its potential for preheating in the heat exchanger. The thermodynamical process of Figure 2.41 is illustrated in Figure 2.42.

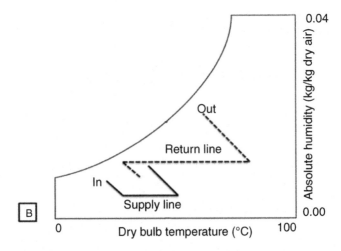

Figure 2.42 Psychrometrics chart of the solid desiccant cooler cycle.

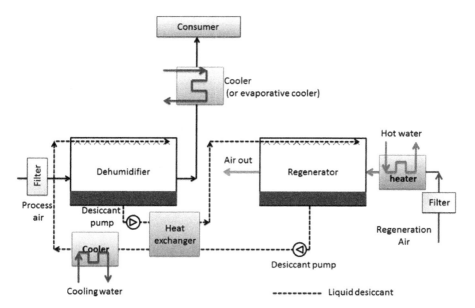

Figure 2.43 A liquid desiccant cooler cycle.

The preheated air is further heated in the heater for use for regeneration of the desiccant. The required heat for heating can be provided by the recoverable waste energy of the prime mover of a CCHP system.

Frequently used solid desiccants include silica gel with a regeneration temperature of 60 to 120 °C, lithium chloride with a regeneration temperature of 80 to 150 °C, calcium chloride with a regeneration temperature of 95 to 120 °C, zeloite with a regeneration temperature greater than 150 °C, molecular sieves, aluminum oxides, etc. [3, 11, 22].

Some recognized companies that manufacture solid desiccant wheels or cassettes include Munters USA, Munters AB, Nichias, DRI, Klingenburg, ProFlute, Bry-Air, Rotor Source, and NovelAir [11].

Liquid desiccants (Figure 2.43) can be used to control the humidity and temperature of air required for a particular consumer. In comparison with solid desiccants, liquid desiccant systems are less complex; they do not need rotary wheels. Liquid desiccants are usually sprayed in the process/regeneration air flow. In addition, similar to cooling towers, packing materials, fills, or splash bars can be used to increase the contact surface area between the liquid desiccant and process/regeneration air to increase dehumidification/regeneration. Liquid desiccants can provide the same dehumidification as solid desiccants with lower regeneration air temperature. Liquid desiccants produce less pressure drop and remove some pollution from the air as well. Carryover of liquid desiccant with the process/regeneration air is a disadvantage of these systems. Corrosion and crystallization are other common problems for the liquid desiccant systems. Some common liquid desiccants include liquid lithium chloride, calcium chloride brine, and sodium chloride brine.

In both solid/liquid desiccant dehumidifiers the process/regeneration filters play an important role in maintaining the desiccant efficiency, therefore special attention must be paid to them. The filters must be checked, cleaned, or replaced periodically according to the air pollution level and this may be different from place to place and even time to time during a year. Maintenance of filters is more economic than replacing the liquid desiccant. The desiccants usually must be replaced, refilled, or reconditioned according to the application after every 5 to 10 years of operation. Without attention to filters the desiccant life may reduce by 1 to 2 years. Filter clogging has a negative impact on the heater (if it has a burner) due to lack of oxygen. It also has a negative impact on the flow rate of process/regeneration air; due to lack of air regeneration may be incomplete and air flow to the consumer decreases as well [22]. This also increases the pressure drop and increases fan power consumption.

2.3.5 Ejector Cooling Cycle

Ejectors use neither heat nor electricity directly as the running input energy of ejector cooling cycles (ECC). They use a high-pressure motive flow of steam to circulate the ECC working fluid. They are usually installed in conjunction with power cycles and organic Rankine cycles. The ejector is the heart of an ECC and is classified into two types based on the mixing method in the primary nozzle exit [23-25]. The first one is the constant pressure jet ejector (CPJE) and the other one is the constant area jet ejector (CAJE). The CPJE is more suitable for a wide condensing pressure range while the CAJE is better for drawing more mass flow rate [23]. The CPJE provides better performance than the CAJE due to better turbulent mixing [24, 26, 27]. Besides having no moving parts, ECC benefits from lower maintenance expenses and lower capital costs than the compressor [28]. In Figures 2.44 and 2.45, the geometry of a CPJE and its enthalpy-entropy diagram are illustrated. As the picture shows, a motive high-pressure flow is expanded in a nozzle and its pressure drops to draw in the suction flow. The motive and suction flows mix with each other at a constant pressure of M, then the mixture passes through a throat and its pressure increases to P_5. The flow finally diffuses in a diffuser and its pressure

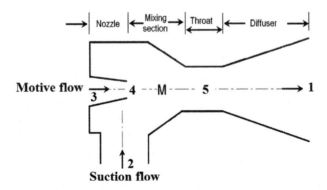

Figure 2.44 The schematic of the CPJE.

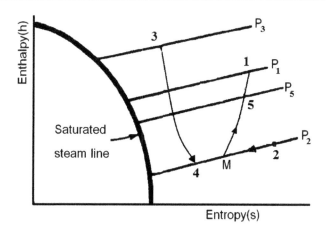

Figure 2.45 Mollier's diagram of the ECC process [18].

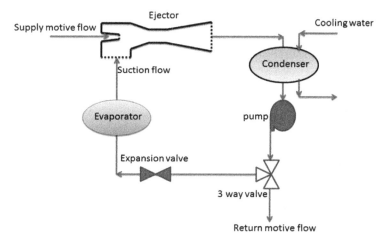

Figure 2.46 Ejector cooling cycle.

reaches P_1 at the exit ejector. The purpose of the ejector is to increase the suction flow pressure from P_2 to P_1.

An ECC is represented in Figure 2.46. The motive flow energy can be provided from the waste energy of different systems.

2.4 Problems

1. According to the information given in this chapter, provide a table that compares all types of prime movers from technical, environmental, and economic points of view.
2. Provide a table of TAC systems that compares them from technical, environmental, and economic points of view.

3. By using the tables provided in Problems 1 and 2, consider the feasibility of different combinations of TAC and prime movers. In this decision-making process what parameters are more important?

4. Draw the thermodynamical cycles for the possible CCHP systems considered in Problem 3.

5. Estimate the capital costs of CCHP components created in Problem 4, if the prime mover size is less than 100 kW.

6. If it is decided to use an auxiliary heating source, what is your recommendation? Do you prefer to use condensing boilers or conventional boilers?

7. If you decide to use a condensing boiler instead of a conventional boiler (in Problem 6), how much different would the capital costs be? What about the boiler efficiency? Does this difference in boiler efficiency pay back the extra money paid in 5 years?

8. If we are to choose the cooling system for a CCHP system, what would you choose among the following three alternatives? Present your discussion from energy and economic points of view.
   a. Using an absorption chiller to satisfy all the cooling demand
   b. Using an electric chiller to satisfy all of the cooling demand
   c. Using a combination of electric and absorption chillers to satisfy all of the cooling demand

9. If the electrical demand of the consumer is changeable, what type of prime mover would you choose according to this variable electrical demand? What are your criteria in this decision-making process?

10. Do ambient temperature/humidity and altitude above sea level change the COP of adsorption chillers and desiccant cooling systems? Propose a method to evaluate the impact.

References

[1] Mapna 22 Yazd, Basic Operation Training, Steam Turbine, Siemens, E8SQ Training Dept, 2005.

[2] COGEN Europe A guide to cogeneration. www.cogen.org. March 2001.

[3] Wu, D.W., Wang, R.Z., 2006. Combined Cooling, Heating and Power: A Review. Progress in Energy and Combustion Science 32, 459–495.

[4] Energy and environmental analysis prepared for Environmental Protection Agency, Combined Heat and Power Partnership Program, Washington DC, December 2008.

[5] www.wolf-energiesparsysteme.de, accessed 2013-12-21.

[6] Onovwiona, H.I., Ugursal, V.I., 2006. Residential Cogeneration Systems: Review of the Current Technology. Renewable and Sustainable Energy Reviews 10, 389–431.

[7] www.capstoneturbine.com, accessed 2013-12-21.

[8] www.energent.net, accessed 2013-12-21.

[9] www.stirling-engine.de, www.cleanergy.com, accessed 2013-12-21.

[10] www.sgc.se, accessed 2013-12-21.

[11] Deng, J., Wang, R.Z., Han, G.Y., 2011. A Review of Thermally Activated Cooling Technologies for Combined Cooling, Heating and Power Systems. Progress in Energy and Combustion Science 37, 172–203.

[12] ASHRAE Handbook–Refrigeration, Absorption Cooling, Heating, and Refrigeration Equipment, Chapter 41, 2002.

[13] Maraver, D., Sin, A., Sebastian, F., Royo, J., 2013. Environmental Assessment of CCHP (Combined Cooling Heating and Power) Systems Based on Biomass Combustion in Comparison to Conventional Generation. Energy 57, 17–23.

[14] www.hitachi-ap.com, accessed 2013-12-21.

[15] www.york.com, accessed 2013-12-21.

[16] www.johnsoncontrols.com, accessed 2013-12-21.

[17] www.rotartica.com, accessed 2013-12-21.

[18] Cengel, Y.A., Boles, M.A., 2006. Thermodynamics: An engineering Approach, Fifth Edition McGraw-Hill. College, Boston, MA, ISBN 10:0072884959

[19] Tichi, S.G., Ardehali, M.M., Nazari, M.E., 2010. Examination of Energy Price Policies in Iran for Optimal Configuration of CHP and CCHP Systems Based on Particle Swarm Optimization Algorithm. Energy Policy 38, 6240–6250.

[20] www.colibri-bv.com, accessed 2013-12-21.

[21] www.mayekawausa.com, accessed 2013-12-21.

[22] ASHRAE Handbook, Desiccant Dehumidification and Pressure-Drying Equipment, Chapter 23, 2008.

[23] Pianthong, K., Seehanam, W., Behnia, M., Sriveerakul, T., Aphornratana, S., 2007. Investigation and Improvement of Ejector Refrigeration System Using Computational Fluid Dynamics Technique. Energy Conversion and Mangement 48, 2556–2564. doi: 10.1016/j.enconman.2007.03.021

[24] Wanavet, S.W., 2005. Optimization of a High-Efficiency Jet Ejector by Computational Fluid Dynamics Software. M. Sc. Thesis, Texas A&M University.

[25] Best, R., October 2007. Recent Developments in Thermal Driven Cooling and Refrigeration Systems. 1st European Conference on Polygeneration. Tarragona (Spain), pp. 16–17.

[26] Defrate, L.A., Hoerl, A.E., 1959. Optimum Design of Ejectors Using Digital Computers. Chem. Eng. Prog. Symp. Series 21.

[27] Kim, H.D., Setoguchi, T., Yu, S., Raghunathan, S., 1999. Navier-Stokes Computations of the Supersonic Ejector-Diffuser System with a Second Throat. Journal of Thermal Science, 2, 79–83.

[28] Ebrahimi, M., Keshavarz, A., Jamali, A., 2012. Energy and Exergy Analyses of a Micro-Steam CCHP Cycle for a Residential Building. Energy and Buildings 45, 202–210.

CCHP Evaluation Criteria

3.1 Introduction

In order to decide upon, design, and optimize a CCHP cycle and its components, the parameters that have impact on these three steps should be analyzed.

In this chapter, different criteria that are commonly used for deciding upon, designing, and optimizing CCHP cycles will be introduced. These criteria consider different aspects of energy and CCHP technology. They include technological, environmental, economic, and miscellaneous criteria. Each of these criteria can be broken into some subcriteria.

The technological subcriteria may include but is not limited to the fuel energy saving ratio (FESR) with respect to the latest, common technologies in the conventional separate production of cooling, heating, and power (SCHP); the overall efficiency of the prime mover and the CCHP cycle; the exergy efficiency (π) of the CCHP system; the performance of the power generation unit (PGU) of the CCHP cycle in partial and full load operation (PLO, FLO); the power to recoverable heat ratio (PHR) of the PGU; the recoverable heat source temperature (T_{rhs}) from the PGU; the user-friendliness of the control and regulation system (UFCR); the maturity of the technologies used in the CCHP cycle; etc.

The environmental subcriteria includes but is not limited to the reduction ratio of air pollutants such as carbon monoxide, carbon dioxide, nitrogen oxides, etc. The noise level also should be considered one of the comfort conditions especially in residential, educational and hospital applications of the CCHP cycles. The ability to use renewable energy resources such as solar heat and biofuel in the CCHP cycle as the main energy source or base energy source (for hybrid systems) also can be discussed in the decision-making process.

The economics of the CCHP cycles is very important for users and investors. Therefore the designer should define the most important economic subcriteria for the project. Among the economic parameters is the net present value (NPV), which guarantees the profitability of the project for the lifetime of the CCHP cycle. The internal rate of return (IRR) is usually used to predict the risk of investment, and the payback period of the project predicts the time needed to recover the original money invested. The other economic parameters that have an impact on the user decision include initial investment cost, annual cash flow, maintenance cost, etc.

The miscellaneous criterion includes some subcriteria that are important but do not fall in any of the previous three categories. For instance, there is the ease of maintenance of the CCHP cycle, the availability of spare parts, the import-export limitations of the CCHP components, the space needed for the CCHP system, the lifetime of the CCHP system, etc.

Some of the criteria that are introduced above are qualitative; therefore they should be converted to quantities. This topic will be discussed in Chapter 4.

Combined Cooling, Heating and Power.

As can be seen there are numerous criteria that have an impact on the decision-making process for the CCHP cycle as well as its design and optimization. Therefore the designer should find a way to make use of the privileges of all of the criteria simultaneously while satisfying technological, economic, and environmental constraints.

3.2 Technological Subcriteria

3.2.1 Fuel Energy Saving Ratio (FESR)

This parameter is used to compare the fuel consumption of a CCHP cycle with the SCHP. For this purpose the fuel consumptions should be calculated to produce the same amount of electricity, cooling, and heating. Therefore we have

$$FESR = \frac{F_{SCHP} - F_{CCHP}}{F_{SCHP}} \tag{3-1}$$

In order to make better judgments about the CCHP cycle, this criterion should be calculated annually. In the above equation F stands for the fuel energy. The calculation of fuel energy for the CCHP and SCHP depends on the prime mover type and other technologies that are used in the CCHP or SCHP. These details will be discussed in Chapter 6 when presenting the design methods of CCHP systems. The FESR is an important parameter in the three steps of decision-making, designing, and optimizing CCHP systems.

3.2.2 Overall Efficiency

The prime mover efficiency (η_{PM}) is defined based on the electricity production (E_{PM}) and the corresponding fuel energy consumption (F_{PM}) of the prime mover at full load operation condition as follows:

$$\eta_{PM} = \frac{E_{PM}}{F_{PM}} \tag{3-2}$$

where the subscript PM stands for the prime mover.

In the decision-making process for the prime mover type, the overall efficiency of the prime mover (η_{o-PM}) should be calculated in order to compare different types of prime mover from this point of view:

$$\eta_{o-PM} = \frac{E_{PM} + Q_{rec}}{F_{PM}} \tag{3-3}$$

where Q_{rec} is the recoverable heat from the prime mover. The quality and magnitude of Q_{rec} varies in different types of prime movers. For example, in a reciprocating engine heat with different qualities can be recovered from the exhaust, water jacketing, and lube oil cooler. In a micro-gas turbine (MGT), the heat can be recovered from the exhaust gas of the turbine with a very high quality. In a fuel cell, the heat is recovered from the heat of chemical reactions and its quality differs from one type of fuel cell to the others.

After selection of the prime mover type, in the design process the overall efficiency of the CCHP cycle is defined as follows. This parameter can also be used in the optimization of the CCHP system:

$$\eta_{o-CCHP} = \frac{E + H + C}{F_E + F_H + F_C} \tag{3-4}$$

In the above equation, E represents the total electricity produced by the prime mover and possible purchase of electricity from the grid. In addition, H represents the total heat consumed by the target building that feeds from the recovered heat, possible solar heater, or auxiliary heating system. Furthermore, C represents the total cooling load of the target building that is provided by the cooling system. The cooling system may use the recovered heat, possible solar heater, or auxiliary boiler and electricity to provide the cooling load. For example, using a combination of absorption and compression chillers enables us to use both the heat and electricity of the CCHP system. The above parameter should be calculated annually to have a better understanding of CCHP cycle performance.

3.2.3 Exergy Efficiency

Exergy is the *work potential* of energy. The thermodynamic second-law efficiency, which is also called exergy efficiency, can be calculated as follows:

$$\pi = \frac{Exergy\ recovered}{Exergy\ supplied} = 1 - \frac{Exergy\ loss}{Exergy\ supplied} \tag{3-5}$$

The above equation can be used for the CCHP cycle and its components as well, where the exergy loss can be determined as follows:

$$\dot{I} = \sum \dot{\phi}_{in} - \sum \dot{\phi}_{out} \tag{3-6}$$

where $\dot{\phi}$ is the exergy rate. The exergy rate of a thermodynamical state of i, heat rate ($\dot{Q}$), and power or work ($\dot{W}$) can be calculated respectively as follows:

$$\dot{\phi}_i = \dot{m}_i[(h_i - h_0) - T_0(s_i - s_0)] \tag{3-7}$$

$$\dot{\phi}_Q = \dot{Q}\left(1 - \frac{T_0}{T_{hs}}\right) \tag{3-8}$$

$$\dot{\phi}_w = \dot{W} \tag{3-9}$$

where the subscripts *in*, *out*, *0*, and *hs* stand for *input*, *output*, thermodynamical dead state ($P_0 = 1$ atm, $T_0 = 25°C$), and *heat source*, respectively.

3.2.4   Power to Heat Ratio (PHR)

This criterion is used while selecting the prime mover type for the CCHP system and is calculated for the target building and the prime mover as well. The PHR of the building and the prime mover can be determined as follows:

$$PHR_{PM} = \frac{E_{nom}}{Q_{rec}} \tag{3-10}$$

$$\overline{PHR}_{building} = \frac{\overline{E_{dem}}}{ATD} \tag{3-11}$$

$$ATD = H_{dem} + C_{dem}/COP_{abc} + D_{dem} \tag{3-12}$$

where E_{nom} is the nominal electrical capacity of the prime mover, $\overline{E_{dem}}$ is the yearly average of electrical demand of the building, and ATD is the aggregated thermal demand of building. In addition, C_{dem}/COP_{abc} is the heat consumed by an absorption chiller with coefficient of performance of COP_{abc} to provide the cooling demand of the building. D_{dem} is the energy required to provide the domestic hot water of the building.

One may ask why the PHR of the building is calculated in the average form, or why we don't calculating it as $PHR_{building, \ max} = E_{dem, \ max}/ATD_{max}$.

To answer this question, we assume a consumer with constant demand of electricity and ATD; therefore for this building $PHR_{building, \ max} = \overline{PHR}_{building}$. From the energy and environmental point of view this is an ideal case, because a prime mover with $E_{nom} = E_{dem}$ and the same PHR as the building will fulfill the building's energy demands, and there is no need for an auxiliary heating system or purchasing electricity from the grid; furthermore no recoverable energy would be wasted. However, in real cases the building's energy demands are not constant; they usually fluctuate. For this reason considering a prime mover with higher PHR as $PHR_{building,max}$ results in wasting more recoverable heat whenever the building demand is less than the recoverable heat. It also requires a connection to the grid for possible sale of electricity or operation of the prime mover in partial load. A prime mover with a PHR much smaller than $PHR_{building}$ results in a larger auxiliary heating system and a connection to the grid for purchasing a significant amount of electricity, which in general means not making use of the CCHP system privileges.

Consequently, considering a prime mover with a PHR closer to the $\overline{PHR}_{building}$ gives us the opportunity to make use of the privileges of CCHP systems and a moderate engine size and auxiliary heating system, which is very important in decreasing the initial investment cost for investors.

3.2.5   Operation in Partial Load of Prime Mover (OPL)

Good operation in partial load means the prime mover nominal efficiency does not experience a significant drop when the load decreases. This characteristic is different in different types of prime movers. For example, the nominal efficiency of a gas turbine or MGT is more sensitive to load changes than internal combustion engines. A good operation in partial load causes the η_{PM} and PHR_{PM} to decrease less, and as a result the overall efficiency of the prime mover stays approximately the same. The electrical demand of most of the buildings is changeable during every day, hour, and even minute or second. Therefore if a prime mover is supposed to work in such conditions, it should operate well in partial load. The efficiency, overall efficiency, and PHR of the prime mover are linked as follows:

$$\eta_{o-PM} = \eta_{PM} \left(1 + \frac{1}{PHR_{PM}} \right) \qquad (3\text{-}13)$$

OPL is a qualitative criterion that is only used in decision-making about the prime mover of the CCHP system. This parameter is quantified by using some techniques in Chapter 4.

3.2.6 User-Friendliness of Control and Regulation (UFCR)

A system that is supposed to be installed in a residential, hospital, educational, or commercial building should have a simple control and regulation system for nonexpert users. As an example, control and regulation of internal combustion engines is simpler than a high-tech MGT or fuel cell. The control system of an MGT with a rotational speed of higher than 10^5 rpm and its auxiliary systems, such as lube oil and air bearings, is more complex than common internal combustion engines. The UFCR is also a qualitative criterion for choosing CCHP components that would be quantified.

3.2.7 Maturity of the Technology

The more mature the technology of the CCHP components the better. Maturity usually results in higher safety of operation, availability of spare parts in many producer companies and maintenance shops, and lower final cost due to mass production in many companies. Maintenance of mature technologies costs less than nonmature technologies, because their technology is well-known for operators and troubleshooters. For example, internal combustion engine technology is more mature than MGT technology, its price per kW is cheaper, and more companies produce combustion engines than MGTs. Although internal combustion engines needs more maintenance but the maintenance job is easier than MGTs. The maturity is also a qualitative criterion for choosing CCHP components that would be quantified.

3.2.8 Recoverable Heat Quality

The quality of the recoverable heat from the prime mover is very important. Heat with low quality has less of an opportunity to be reused. The temperature of recoverable heat is different in various prime mover technologies. For example, in internal combustion engines, the lube oil, water jacketing, and exhaust gases have temperatures of approximately about 70°C, 85–95°C, and 370–540°C, respectively. The exhaust temperature of an MGT, Stirling engine, and different types of fuel cells ranges from 200–350°C, 60–300°C, and 50–1000°C, respectively. These temperature ranges are especially important when choosing thermally activated technologies such as single-, double-, or triple-effect absorption chillers, adsorption chillers, dehumidification, etc.

3.3 Economic Subcriteria

3.3.1 Initial Investment Cost (I)

The initial cost of a CCHP system is the first economic aspect that a user or buyer faces. This parameter alone does not show the profitability, risk of investment, or payback period of the project, but it has a significant impact on user decision-making. The lower the initial cost, the more the customer will be interested in using CCHP systems. For this reason, the initial cost is considered as a criterion for selection of CCHP components. In economic design, it is not a design parameter, but has a significant impact in the calculation of all of the economic subcriteria.

3.3.2 Operation and Maintenance Cost (I_{OM})

The second economic subcriterion that has a significant impact is the operation and maintenance cost of the CCHP system. Paying high costs for both regular and unexpected minor and major maintenance and overhaul has a negative impact on the decision-making of the consumer. For instance, internal combustion engines have a higher I_{OM} due to having reciprocating motion and pulse generation in the system. On the contrary, dynamic motion prime movers such as MGTs work continuously without pulse generation; therefore one of the most common reasons for maintenance problems would be omitted naturally in MGTs. This parameter can be considered as a qualitative or quantitative criterion, because the maintenance cost for most stationary and rotary equipment is known. This criterion is considered in decision-making as an individual parameter, but in the design process it is included in the annual expenses of the CCHP system.

3.3.3 Cash Flow (cf)

Cash flow is the net annual profit from the project. The higher the cash flow is, the more interest there will be in using CCHP systems. Cash flow can be calculated by subtracting the expenses (ex) from the earnings (er) for each time step. This parameter should be calculated for every year of the lifetime (L) of the project:

$$cf_y = \sum_{t=1}^{Y}(er - ex)_t, \quad y = 1, 2, ..., L \qquad (3\text{-}14)$$

where Y is the number of time steps during a year, which is usually assumed to be 8760 hours. It should be noted here that every omitted cost in the CCHP with respect to the SCHP and the money that is received due to selling electricity should also be included in the earnings. The money that is paid for buying fuel, electricity, maintenance, etc. is counted as expenses. This parameter is also important for decision-making, because when a project has a high cash flow it increases the interest of customers in using this technology. The positive impact of high cash flow may decrease the negative effect of the high initial investment cost, because the customer can easily compute the payback period of the project by dividing the initial investment cost by the cash flow.

3.3.4 Payback Period (PB)

This criterion can be considered a rule of thumb; it tells how long it takes to recover the original investment I and is calculated as follows:

$$PB = \frac{I}{\overline{cf}} \tag{3-15}$$

$\overline{cf}$ is the average cash flow during the lifetime of the project. As can be seen this criterion does not consider the real value of the money as time passes (in other words it does not consider the inflation rate); this is the weakness of this criterion. PB is simple and understandable for nonexperts. The CCHP cycle and its components can be designed to achieve a predefined PB_{max}. PB is considered as one of the designing parameters for economic evaluations of the CCHP project. The unit of PB is year.

3.3.5 Net Present Value (NPV)

NPV is a strong criterion to determine if a project is profitable or not. It considers the interest rate (r), which is usually equal to the inflation rate; therefore the real value of money at every year of operation is considered. NPV is calculated as follows:

$$NPV = \sum_{y=1}^{L} \frac{cf_y}{(1+r)^y} - I \tag{3-16}$$

For a project to be profitable, NPV must be positive. The more positive the NPV is, the more profit the project will produce. The CCHP cycle and its components can be designed to achieve a predefined profit of NPV_{min} after L years. The unit of NPV is the unit of the currency used in the economic calculations.

3.3.6 Internal Rate of Return (IRR)

The calculation of IRR is vital for the economic evaluation of CCHP systems, because it calculates the profitability margin ($IRR - r$) of the project. In fact, the IRR calculates the interest rate for which the NPV of the project would be zero. If the IRR becomes smaller than r, investment in the project would lose money, for IRR = r the profit is zero and there is a risk of losing money, and for IRR > r the investment is safe. IRR can be calculated according to the following equation:

$$NPV\big|_{IRR} = \sum_{y=1}^{L} \frac{cf_y}{(1+IRR)^y} - I = 0 \qquad (3\text{-}17)$$

To guarantee the investment, the CCHP cycle and its components can be designed to have a predefined $IRR = IRR_{min} > r$.

3.4   Environmental Subcriteria

3.4.1   Noise

According to the application of CCHP systems, different noise levels are defined as the comfort condition. Noise is especially important for residential, hospital, and school CCHP users. For example, an acceptable noise level in the bedroom at night is 30 db, while acceptable noise levels for living areas during the day, school classrooms during the day, and hospital patient rooms during day and night are 50–55 db, 35 db, and 30–35 db, respectively. This parameter is used during component selection for the CCHP system.

3.4.2   Air Pollution

Depending on the type of fossil fuel burned in a CCHP system, different combustion products such as NO_x, CO, CO_2, unburned hydrocarbons, volatile organic compounds, SO_x, and particulate matter (PM) may be released into the atmosphere. The amount of fuel consumption has a direct impact on the amount of air pollution produced, therefore decreasing fuel consumption means decreasing air pollution. As the prime mover size decreases, the nominal efficiency of the prime mover decreases as well. This means more fuel consumption to produce the same amount of electricity in comparison with larger power plants with the same type of power generation unit. In the small CCHP units with which this text is concerned, the waste heat of the prime mover is recovered for reuse to provide for the cooling and heating demands of the consumer. This recovery creates a fuel consumption reduction that compensates for the increase in fuel consumption for electricity production. Calculations that will be presented in the following chapters show that annual savings of fuel for CCHP systems are more than 20%.

Another parameter that has a significant impact on emission reduction is the technology of the combustion chamber and pollution control system of the prime mover and cooling and heating units. Different methods are used to reduce pollutants; among them the following techniques can be mentioned: combustion process emissions control, post-combustion emissions control, three-way catalyst, selective catalytic reduction (SCR), oxidation catalysts, and lean-NO_x catalysts.

Therefore, CCHP systems with emission reduction systems considerably reduce air pollution in addition to the reduction in fuel consumption.

The emission production index of different technologies is usually given as i (kg/MWh), so that the mass of yearly emission production ($Em(g)$) due to providing x kW of heating or electricity can be calculated as follows:

$$Em_X = \sum_{t=1}^{L} (x \cdot i)_t \qquad\qquad (3\text{-}18)$$

where X is the emission type such as CO, CO_2, etc.

3.5 Miscellaneous Subcriteria

3.5.1 Import and Export Limitations (IEL)

Depending on the country that the CCHP is designed for, there may be some social, political, economic, technological, environmental, etc. limitations on using, importing, or exporting the CCHP components; therefore at the decision-making stage attention must be paid to this matter.

3.5.2 Local Ability and Ease of Maintenance (LAEM)

If local operators or staff are able to do the maintenance of the CCHP components it will decrease the shutdown time of the cycle. In addition CCHP systems with simple maintenance jobs consume less time for overhaul and unscheduled shutdowns. The purpose of a CCHP system is to provide all or a considerable amount of the energy demands of a building during a year. Therefore highly available and reliable CCHP systems are needed, especially for residential, hospital, school, and commercial buildings. Any unwanted cutoff due to minor or major maintenance problems increases the downtime of the unit and decreases the reliability of the CCHP system. Availability is related to the scheduled downtime (S) while reliability is related to the unscheduled downtime (U) due to component failure of the CCHP system; they can be calculated as follows:

$$Reliability\ (\%) = \frac{T-(S+U)}{T-S} \times 100 \qquad\qquad (3\text{-}19)$$

$$Availability\ (\%) = \frac{T-(S+U)}{T} \times 100 \qquad\qquad (3\text{-}20)$$

where T is the number of hours that the CCHP needs to be in service per year. The unit of S and U is hours per year. Preventive maintenance and annual overhaul are done during the scheduled downtime hours. The availability of CCHP systems is usually about 95% to 98%.

The local ability and ease of maintenance have a great impact on the reliability and popularity of CCHP systems as well.

3.5.3 Lifetime

A long lifetime for a CCHP system has positive economical effects; it increases the total cash flow and NPV, and decreases the risk of investment. It also has a positive impact on the judgment of a customer for decision-making.

3.5.4 Footprint

The area that the installed CCHP occupies is also considered in the decision-making step.

3.6 Problems

1. According to the literature review and technologies presented in Chapters 1 and 2, introduce some other evaluation criteria for the CCHP systems.
2. Classify the criteria you have chosen in Problem 1 into qualitative and quantitative criteria.

References

[1] Wu, D.W., Wang, R.Z., 2006. Combined Cooling, Heating and Power: A Review. Progress in Energy and Combustion Science 32, 459–495.
[2] Onovwiona, H.I., Ugursal, V.I., 2006. Residential Cogeneration Systems: Review of the Current Technology. Renewable and Sustainable Energy Reviews 10, 389–431.
[3] Biezma, M.V., San Cristbal, J.R., 2006. Investment Criteria for the Selection of Cogeneration Plants—A State of the Art Review. Applied Thermal Engineering 26, 583–588.
[4] Wang, J.-J., Jing, Y.-Y., Zhang, C.-F., Zhang, X.-T., Shi, G.-H., 2008. Integrated Evaluation of Distributed Triple-Generation Systems Using Improved Grey Incidence Approach. Energy 33, 1427–1437.
[5] Jing, Y.-Y., Bai, H., Wang, J.-J., 2012. A Fuzzy Multi-Criteria Decision-Making Model for CCHP Systems Driven by Different Energy Sources. Energy Policy 42, 286–296.
[6] Ebrahimi, M., Keshavarz, A., Jamali, A., 2012. Energy and Exergy Analyses of a Micro-Steam CCHP cycle for a Residential Building. Energy and Buildings 45, 202–210.

CCHP Decision-Making

4

4.1 Introduction

Every decision we make about the essential components of the CCHP system, including the power generation unit, the cooling system, and the heat recovery system, changes the quality and quantity of all of the criteria and subcriteria that were discussed in Chapter 3.

As discussed in Chapters 1 and 2, there are several options for each main component of the CCHP system. Available prime movers include different types of industrial gas and steam turbines, micro-gas turbines (MGTs), micro-steam turbines (MSTs), internal combustion engines (compression ignition, spark ignition), different types of external combustion Stirling engines, and fuel cells.

Each of the above technologies can be coupled with photovoltaic (PV) cells or electricity storage systems to provide or store extra electricity to fulfill the electrical demand of the building or sell it during electrical peak hours at higher prices.

The heat in the CCHP systems is recovered, generated, or stored for three purposes. Heat is used for heating spaces via terminal stations such as fan coil units on cold days, cooling spaces via absorption or adsorption chillers on hot days, and providing the domestic hot water demand of the target building during the year.

The domestic hot water system can be either integrated with or separated from the CCHP system. When it is separated from the CCHP system, the hot water can be provided by using a solar water heater or a gas fired water heater. When it is integrated with the CCHP system, it uses the recoverable heat from the prime mover, auxiliary heating system, or solar water heater.

Calculations must be done to compare the technological, economic, etc. characteristics of the CCHP system to decide whether the domestic hot water should be integrated with or separated from the CCHP system.

Addition of any new component such as an electrical compression chiller, solar PV cells, electricity storage system, solar water heater, auxiliary heating system, or heat storage system to the basic CCHP unit changes the CCHP's characteristics. Comprehensive evaluations must be done prior to decision-making and the addition of any new component.

According to the discussion presented above, decision-making can be divided into two levels:

1. Basic decision-making
2. Design decision-making

In the *basic decision-making* process (which is the main concern of this chapter), only the basic components (the prime mover, cooling system, and heating system) are decided (Figure 4.1). For this purpose multicriteria decision-making (MCDM) methods will be introduced.

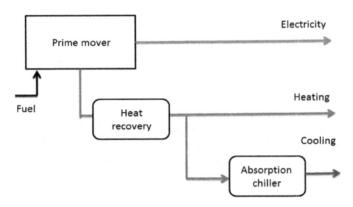

Figure 4.1 A CCHP unit with basic components.

However, in the *design decision-making* process (which is a part of detail design and will be presented in the following chapters), decisions are made about the presence, position and size of the components in the CCHP cycle, (Figure 4.2).

Different and numerous criteria are encountered during the decision-making process for the basic components of a CCHP system. The most important criteria involved in the decision-making process are presented in the previous chapter. They have the following characteristics:

- Some of the criteria are qualitative (maturity, UFCR, OPL, LAEM).
- The criteria have different dimensions (NPV ($), noise (db), PB (year)).
- The criteria have different orders of magnitude ($0 < FESR < 1$, NPV can be as high as 10^5).

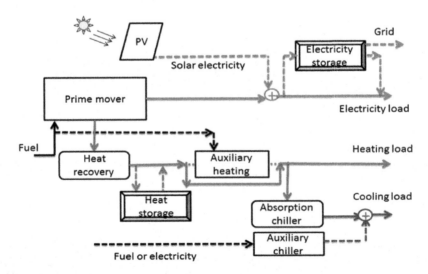

Figure 4.2 Schematic of a CCHP unit with some nonbasic components.

- The criteria have different levels of importance (FESR is more important than OPL or PHR, …).
- There are numerous important criteria (FESR, NPV, IRR, IEL, …).

Decision-making based on the above criteria with all their different characteristics is very difficult. MCDM must be able to compromise between all different criteria, and for this purpose it must have the following abilities:

1. It must make all of the qualitative criteria quantitative.
2. It must present all the criteria in dimensionless form.
3. It must normalize all the criteria (they all should be in the same range)
4. It must give normalized weight to each criterion and subcriterion according to the available technical data and knowledge of the decision-maker.
5. It must integrate all of the criteria into one or a limited number of criteria to make the decision-making easier.

4.2 Multicriteria Decision-Making (MCDM)

Probability and statistics, fuzzy mathematics, and grey systems theory are the three most commonly used research methods employed for the investigation of uncertain systems [1]. Decision-making for CCHP components includes uncertain parameters such as maturity, PHR, OPL, LAEM, IEL, UFCR, etc. They are uncertain parameters because they are judged with linguistic concepts such as very bad, bad, moderate, good, and very good, or very low, low, medium, high, and very high. Uncertainty in these cases means that we cannot define a clear border between two neighboring qualities such as bad and very bad.

Probability and statistics studies the chance of a phenomena with an uncertain nature happening in the future based on a large statistical sample data from the past. Probability calculations start with large sample data to generate a typical statistical distribution such as normal, uniform, etc.

Zadeh (1965) [2] was the first to propose the fuzzy set. The fuzzy set includes elements with only a partial degree of membership. In recent years, the applications of fuzzy logic have increased significantly. The applications range from products such as cameras, washing machines, and microwave ovens to industrial process control and decision-making systems. Fuzzy logic starts with the concept of a fuzzy set. A fuzzy set is a set without a crisp, clearly defined boundary. It can contain elements with only a partial degree of membership. In the following three different membership functions, namely the triangular-shaped, generalized bell-shaped, and symmetric Gaussian functions, are presented. The triangular-shaped membership function can be written as [3]

$$f_x(x, a, b, c) = \max\left(\min\left(\frac{x-a}{b-a}, \frac{c-x}{c-b} \right), 0 \right) \tag{4-1}$$

while the generalized bell-shaped membership function can be written as

$$f_x(x, a, b, c) = \left(1 + \left|\frac{x-c}{a}\right|^{2b}\right)^{-1} \tag{4-2}$$

In the preceding functions (a, b, c) is a triangular fuzzy number.

Finally, the symmetric Gaussian membership function, which depends on two parameters (σ, c) can be written as follows:

$$f_x(x, \sigma, c) = e^{\frac{-(x-c)^2}{2\sigma^2}} \tag{4-3}$$

The behavior of the preceding membership functions is depicted in Figure 4.3.

If $I = (a_I, b_I, c_I)$ and $II = (a_{II}, b_{II}, c_{II})$ are two triangular fuzzy numbers, their summation, multiplication, and division are as follows:

$$\begin{aligned}
I + II &= (a_I + a_{II}, b_I + b_{II}, c_I + c_{II}) \\
I \times II &= (a_I \times a_{II}, b_I \times b_{II}, c_I \times c_{II}) \\
I / II &= (a_I / c_{II}, b_I / b_{II}, c_I / a_{II})
\end{aligned} \tag{4-4}$$

Grey systems theory was introduced by Deng Ju-long (1982) in China. The main capability of grey systems theory is the ability to study uncertainty problems with very few points of sample data and/or poor information; these problems are difficult with

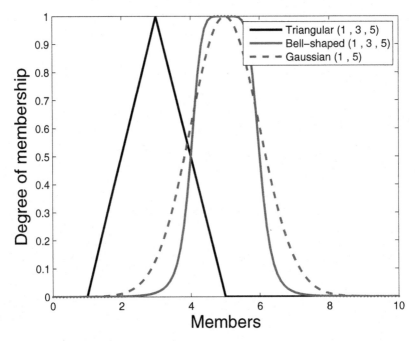

Figure 4.3 Three membership functions.

probability and the fuzzy algorithm. The grey incidence approach (GIA) is one of the main results of grey systems theory. The main idea of the GIA is that the closeness of a relationship is judged based on the degree of similarity of the geometrical patterns of sequence curves [1]. The more similar the curves are, the higher grey incidence grad (GIG) and as a result the better decision to make, and vice versa.

In the following, the two most practical techniques of the fuzzy algorithm and GIA will be introduced and applied to a similar case study to compare their results.

4.2.1 Fuzzy-MCDM Modeling

The fuzzy-MCDM method integrates numerous and different criteria to make the decision more comprehensive. In this section, the corresponding steps of the fuzzy-MCDM method are presented sequentially.

4.2.1.1 Quantifying the Qualitative Criteria

In decision-making problems, some of the criteria such as OPL, maturity, UFCR, IEL, and LAEM are qualitative and it is not easy to quantify them. The importance of a qualitative criterion depends on the technical data, and the opinion and knowledge of the decision-maker. Therefore, fuzzy-MCDM must propose a method to quantify qualities such as "very bad," "bad," "middle," "good," and "very good." In the fuzzy-MCDM, a linguistic judgment is used and the corresponding triangular fuzzy numbers proposed by [4] are used to quantify the qualitative criteria. The linguistic judgments and their corresponding fuzzy numbers are presented in Table 4.1 [5, 6].

In what follows the uppercase letters show the fuzzy numbers and the lowercase letters indicate the real numbers.

4.2.1.2 Generating the Judgment Matrix

Fuzzy decision-making for the basic components of a CCHP system starts by generating a judgment matrix as follows:

$$J = \left[J_{ij} \right]_{m \times n} \tag{4-5}$$

where J_{ij} is our judgment about the ith option based on the jth criteria, with $i = 1, 2...,$ m and $j = 1, 2..., n$. In addition, m and n are the number of alternatives for each component and the number of criteria, respectively.

Table 4.1 Linguistic Judgments and Their Corresponding Fuzzy Numbers

Linguistic Judgment	Fuzzy Judgment
Very low (VL)	(0, 0, 0.3)
Low (L)	(0, 0.3, 0.5)
Middle (M)	(0.3, 0.5, 0.7)
High (H)	(0.5, 0.7, 1.0)
Very high (VH)	(0.7, 1.0, 1.0)

4.2.1.3 Normalization of the Judgment Matrix

To allow criteria to be properly evaluated while comparing options, all the criteria must fall in the same range. For this purpose, the judgment matrix is normalized.

If the criterion is "the higher the better" such as the FESR then

$$J_{ij}^{N} = \frac{J_{ij}}{J_{j}^{\uparrow}} \tag{4-6}$$

If the criterion is "the lower the better" such as the investment cost then

$$J_{ij}^{N} = \frac{J_{j}^{\downarrow}}{J_{ij}} \tag{4-7}$$

where in the preceding equations

$$\begin{aligned} J_{j}^{\uparrow} &= \max\{J_{ij} \mid i = 1, 2 \dots, m\} \\ J_{j}^{\downarrow} &= \min\{J_{ij} \mid i = 1, 2 \dots, m\} \end{aligned} \tag{4-8}$$

If in the normalization process a component of the triangular fuzzy number becomes greater than 1, it will be substituted with 1. Therefore, the normalized judgment matrix (J^{N}) would be as follows:

$$J^{N} = \left[J_{ij}^{N} \right]_{m \times n} \tag{4-9}$$

4.2.1.4 Calculating the Normalized Weight Matrix

In MCDM methods, the degree of importance of each criterion with respect to the other criteria is considered by calculating the weight of each criterion. For this purpose, the pairwise comparison (PC) matrix between the n criteria is made as follows:

$$PC = \left[PC_{ij} \right]_{n \times n} \tag{4-10}$$

where PC_{ij} represents the importance of the ith criterion with respect to the jth criterion. In the fuzzy-MCDM, a linguistic pairwise comparison presented by [4] is used to provide the PC matrix as in Table 4.2.

The equation to calculate the weight of each criterion or subcriterion in the fuzzy eigenvector of the PC matrix is

$$E_{i} = \left[\prod_{j=1}^{n} PC_{ij} \right]^{1/n} \tag{4-11}$$

Table 4.2 Fuzzy Linguistic Pairwise Comparison and Their Corresponding Fuzzy Numbers

Linguistic Judgment	Fuzzy-PC	Fuzzy-PC(R)
Just equal (JE)	(1,1,1)	(1,1,1)
Equally important (EI)	(1/2,1,3/2)	(2/3,1,2)
Weakly more important (WMI)	(1,3/2,2)	(1/2,2/3,1)
Strongly more important (SMI)	(3/2,2,5/2)	(2/5,1/2,2/3)
Very strongly more important (VSMI)	(2,5/2,3)	(1/3,2/5,1/2)
Absolutely more important (AMI)	(5/2,3,7/2)	(2/7,1/3,2/5)

Finally, the normalized weight matrix for each criterion and subcriterion would be calculated as follows:

$$W_i = \frac{E_i}{\sum\limits_{i=1}^{n} E_i} \tag{4-12}$$

4.2.1.5 Finding the Ideal and Anti-ideal Solutions for All Criteria

The case in which all the subcriteria are the best simultaneously is the ideal solution $(I^\uparrow)$ and the case in which all the criteria are the worst simultaneously would be the anti-ideal solution $(I^\downarrow)$. Therefore, the best and the worst solutions would be as follows:

$$I^\uparrow = \left[I_1^\uparrow, I_2^\uparrow \dots, I_n^\uparrow \right]$$
$$I_j^\uparrow = \max\{ J_{ij}^N \big| i = 1, 2, \dots, m \} \tag{4-13}$$

$$I^\downarrow = \left[I_1^\downarrow, I_2^\downarrow \dots, I_n^\downarrow \right]$$
$$I_j^\downarrow = \min\{ J_{ij}^N \big| i = 1, 2, \dots, m \} \tag{4-14}$$

4.2.1.6 Finding the Weighted Distance from the Ideal and Anti-ideal Solutions

In this step, we calculate the weighted distance. The distance from the ideal and anti-ideal solutions is calculated as follows:

$$d_{ij}^{\uparrow\downarrow}(I_j^{\uparrow\downarrow}, J_{ij}^N) = \sqrt{\frac{1}{4}((a_j^{\uparrow\downarrow} - a_{ij}^N)^2 + 2(b_j^{\uparrow\downarrow} - b_{ij}^N)^2 + (c_j^{\uparrow\downarrow} - c_{ij}^N)^2)} \tag{4-15}$$

The weighted distance can be calculated according to the following equation:

$$D_i^{\uparrow\downarrow} = \sum_{j=1}^{n} \beta_j W_j d_{ij}^{\uparrow\downarrow}, \quad i = 1, 2, \ldots, m \tag{4-16}$$

where $\beta \in \{0, 1\}$ is the criteria coefficient and is used for the purpose of single- or multicriteria decision-making. When a criterion is supposed to be counted in the decision-making process $\beta = 1$, otherwise $\beta = 0$.

4.2.1.7 Calculating the Closeness Number (CN)

The closeness of different options to the ideal solution is calculated according to the following equations:

$$\begin{aligned} ND^{\downarrow} &= \min\{D_i^{\downarrow} i = 1, 2, \ldots, m\}, \quad ND^{\uparrow} = \max\{D_i^{\downarrow} i = 1, 2, \ldots, m\} \\ PD^{\downarrow} &= \min\{D_i^{\uparrow} i = 1, 2, \ldots, m\}, \quad PD^{\uparrow} = \max\{D_i^{\uparrow} i = 1, 2, \ldots, m\} \end{aligned} \tag{4-17}$$

$$\alpha_i^{\uparrow\downarrow} = d(D_i^{\uparrow\downarrow}, PD^{\downarrow\uparrow}) + d(D_i^{\downarrow\uparrow}, ND^{\uparrow\downarrow}) \tag{4-18}$$

Based on this, the closeness number (CN_i) is used in the final decision-making process and is calculated as follows:

$$CN_i = \frac{\alpha_i^{\downarrow}}{\alpha_i^{\downarrow} + \alpha_i^{\uparrow}}, \quad i = 1, 2, \ldots, m \tag{4-19}$$

It is obvious that $0 \le CN_i \le 1$, therefore as CN_i approaches 1 it is getting closer to the ideal solution; on the contrary if CN_i approaches zero it is getting closer to the anti-ideal solution.

4.2.2 Grey-MCDM Modeling

4.2.2.1 Quantifying the Qualitative Criteria

In order to quantify the qualitative criteria, in the GIA, a letter grade from A to E is utilized, in which A represents the *very good* and E represents the *very bad*. Moreover, the scale shown in Figure 4.4 is used to quantify these letter judgments [7-9].

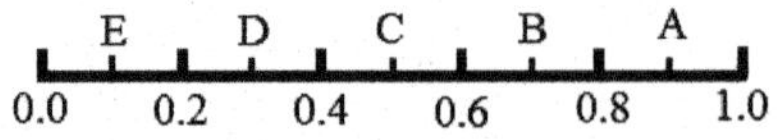

Figure 4.4 The scale for quantifying the qualitative criteria in the GIA.

4.2.2.2 Generating the Judgment Matrix

The grey-MCDM starts by writing the judgment matrix as follows:

$$J = \left\lfloor J_{ij} \right\rfloor_{m \times n} \tag{4-20}$$

where J_{ij} is our judgment about the ith option based on the jth criteria with $i = 1, 2...,$ m and $j = 1, 2..., n$. In addition, m and n are the number of alternatives for each component and the number of criteria, respectively.

For example, if the problem is to find the best type of prime mover, and there are 5 types of prime movers and 15 subcriteria, $m = 5$ and $n = 15$. If the third type of prime mover is an MGT and the first subcriteria is the FESR, then $J_{31} = FESR_{MGT}$.

4.2.2.3 Normalization of the Judgment Matrix

In order to normalize the judgment matrix from the previous section, if the criterion (J_{ij}) is a profit for the project and is "the higher the better" such as the UFCR, then it is normalized as below:

$$J_{ij}^N = \frac{J_{ij} - J_j^\downarrow}{J_j^\uparrow - J_j^\downarrow} \tag{4-21}$$

If the J_{ij} is a cost for the project and is "the lower the better" such as I_{OM}, then it is normalized as follows:

$$J_{ij}^N = \frac{J_j^\uparrow - J_{ij}}{J_j^\uparrow - J_j^\downarrow} \tag{4-22}$$

where in the above equations

$$\begin{aligned} J_j^\uparrow &= \max\{j_{ij} \mid i = 1, 2..., m\} \\ J_j^\downarrow &= \min\{j_{ij} \mid i = 1, 2..., m\} \end{aligned} \tag{4-23}$$

Therefore, the normalized judgment matrix (J^N) would be as follows:

$$J^N = \left[J_{ij}^N \right]_{m \times n} \tag{4-24}$$

4.2.2.4 Calculating the Normalized Weight Matrix

To calculate the normalized weight matrix in the GIA an equation is recommended to determine the pairwise comparison (PC) matrix as follows:

$$PC_{ij} = 1 \pm 0.1\delta, \quad \delta = 0, 1, 2, ..., 10 \tag{4-25}$$

Table 4.3 The Quantified Grey-MCDM Linguistic Pairwise Comparison [5]

Linguistic Judgment	Grey-PC	Grey-PC(R)
Just equal (JE) ($\delta = 0$)	1	1
Equally important (EI) ($\delta = 2$)	1.2	0.8
Weakly more important (WMI) ($\delta = 4$)	1.4	0.6
Strongly more important (SMI) ($\delta = 6$)	1.6	0.4
Very strongly more important (VSMI) ($\delta = 8$)	1.8	0.2
Absolutely more important (AMI) ($\delta = 9$)*	1.9	0.1

* If $\delta = 10$, it means that a criterion has no importance at all with respect to the other criterion, which is not true, because all the criteria have a minimum degree of importance; therefore in this case $\delta = 9$ is more logical.

where the plus sign is for the time when the ith criteria is δ degrees more important than the jth criteria, while the minus sign is for the time when the jth criteria is δ degrees more important than the ith criteria.

According to the decision-maker's opinion and the preceding equation, Table 4.3 is recommended for the linguistic pairwise comparison in the grey-MCDM.

However, the improved GIA proposed by [7] uses a combined weighting method. This method uses an objective weighting method called the entropy information method (EIM) together with a subjective weighting method called the analytic hierarchy process (AHP) to make use of the advantages of both methods. The AHP is mostly based on the knowledge and opinion of the decision-maker while the EIM is based on the technical data and judgment matrix. The combined weighting method blends these two weights together.

The subjective weighting method (AHP) is briefly presented in the following:

$$\rho_j = \sum_{i=1}^{n} PC_{ij}$$
$$\rho_{max} = \max(\rho_1, \rho_2, \ldots, \rho_n)$$
$$\rho_{min} = \min(\rho_1, \rho_2, \ldots, \rho_n)$$
$$\theta = \rho_{max}/\rho_{min}$$
(4-26)

$$b_{ik} = \begin{cases} 1+(\rho_i - \rho_k)/\rho_{min}, & \rho_i \geq \rho_k, \theta \neq 1 \\ (1+(\rho_i - \rho_k)/\rho_{min})^{-1}, & \rho_i < \rho_k, \theta \neq 1 \\ 0 & \theta = 1 \end{cases}$$
(4-27)

$$c_{ik} = \log b_{ik}, \quad i,k = 1, 2, \ldots, n$$
(4-28)

$$d_{ik} = \frac{1}{n}\sum_{l=1}^{n}(c_{il} - c_{kl})$$
(4-29)

$$b'_{ik} = 10^{d_{ik}}$$
(4-30)

Therefore, the characteristic vector of b'_{ik} is given as follows:

$$E_i = \sqrt[n]{\prod_{k=1}^{n} b'_{ik}}, \quad i = 1, 2, \ldots, n \tag{4-31}$$

Finally, the normalized weight matrix in the AHP method would be as follows:

$$W_i^1 = \frac{E_i}{\sum_{i=1}^{n} E_i} \tag{4-32}$$

$$0 \le W_i^1 \le 1, \sum_{i=1}^{n} W_i^1 = 1$$

The above equation gives the normalized weight of each criterion and subcriterion in the AHP.

To compute the combined weight in the GIA, the EIM is used as the objective weighting method. In the EIM method, the proximity degree (PD) of benefit and cost criteria is calculated as follows:

$$PD_{ij} = \frac{J_{ij}}{J_j^{\uparrow}}, \text{ for benefits} \tag{4-33}$$

$$PD_{ij} = \frac{J_j^{\downarrow}}{J_{ij}}, \text{ for costs} \tag{4-34}$$

Then the PD should be normalized as follows:

$$PD_{ij}^N = \frac{PD_{ij}}{\sum_{j=1}^{m} PD_{ij}}, \quad \sum_{j=1}^{m} PD_{ij}^N = 1 \tag{4-35}$$

The entropy of information shows the uncertainty of the criteria and can be calculated as follows:

$$S_i = -\sum_{j=1}^{m} PD_{ij}^N \ln PD_{ij}^N \tag{4-36}$$

The entropy should be normalized with the maximum entropy. According to the previous equation the maximum entropy occurs when all the elements of PD_{ij}^N become equal; therefore we have

$$
\begin{aligned}
S_{max} &= \ln m \\
S_i^N &= S_i / S_{max} \\
S &= \sum_{i=1}^{n} S_i^N
\end{aligned}
\tag{4-37}
$$

Finally, the weights proposed by the EIM can be calculated:

$$
\begin{aligned}
W_i^2 &= \frac{1}{n-S}(1-S_i^N) \\
0 &\le W_i^2 \le 1, \sum_{i=1}^{n} W_i^2 = 1
\end{aligned}
\tag{4-38}
$$

Up to this point, the weights of the criteria and subcriteria have been calculated using the AHP and EIM; now a linear combination is used to calculate the combined weight (W^C) used in the GIA:

$$
W_i^C = \sum_{k=1}^{2} \lambda_k W_i^k, \sum_{k=1}^{2} \lambda_k = 1
\tag{4-39}
$$

In the above equation, $k = 1$ and $k = 2$ correspond to the weights calculated by the AHP and EIM, respectively. The coefficients λ_1 and λ_2 are calculated as follows:

$$
(\lambda_1, \lambda_2) = \left(\frac{S_1}{\sum_{k=1}^{2} S_k}, \frac{S_2}{\sum_{k=1}^{2} S_k} \right)
\tag{4-40}
$$

$$
s_k = \exp\left\{ -\left[1 + \frac{\mu \sum_{i=1}^{n} \sum_{j=1}^{m} W_i^k (1-J_{ij}^N)}{1-\mu} \right] \right\}, \quad k = 1, 2
\tag{4-41}
$$

where μ is the balance coefficient and $0 < \mu < 1$. The impact of μ on the λ's and on decision-making will be presented in the case studies at the end this chapter.

4.2.2.5   Finding the Ideal and Anti-ideal Solutions for All Criteria

The case in which all the criteria are the best simultaneously is the ideal solution ($I^\uparrow$) and the case in which all the criteria are the worst simultaneously would be the anti-ideal solution ($I^\downarrow$). Therefore, the best and the worst solutions for the GIA are as follows:

$$I^\uparrow = \left[I_1^\uparrow, I_2^\uparrow ..., I_n^\uparrow \right]$$
$$I_j^\uparrow = \max\{J_{ij}^N \mid i = 1, 2, ..., m\} \tag{4-42}$$

$$I^\downarrow = \left[I_1^\downarrow, I_2^\downarrow ..., I_n^\downarrow \right]$$
$$I_j^\downarrow = \min\{J_{ij}^N \mid i = 1, 2, ..., m\} \tag{4-43}$$

4.2.2.6   Finding the Weighted Distance from the Ideal and Anti-ideal Solutions

The incidence coefficient between the comparative sequence and the ideal and anti-ideal solutions is calculated as

$$d_{ij}^{\uparrow\downarrow}(I_i^{\uparrow\downarrow}, J_{ij}^N) = \frac{\min_j \min_i \left| I_i^{\uparrow\downarrow} - J_{ij}^N \right| + \xi \max_j \max_i \left| I_i^{\uparrow\downarrow} - J_{ij}^N \right|}{\left| I_i^{\uparrow\downarrow} - J_{ij}^N \right| + \xi \max_j \max_i \left| I_i^{\uparrow\downarrow} - J_{ij}^N \right|} \tag{4-44}$$

where the $\min_j \min_i \left| I_i^{\uparrow\downarrow} - J_{ij}^N \right|$ and $\max_j \max_i \left| I_i^{\uparrow\downarrow} - J_{ij}^N \right|$ are the global minimum and maximum of absolute distance between the normalized judgment criteria from the best and worst solutions. In addition, ξ is the distinguishing coefficient, and $0 < \xi < 1$. The impact of the distinguishing coefficient on decision-making is also studied in the case studies at the end of this chapter.

Finally, the weighted distance can be calculated according to the following equation:

$$D_i^{\uparrow\downarrow} = \sum_{j=1}^n \beta_j W_j^c d_{ij}^{\uparrow\downarrow}, \quad i = 1, 2, ..., m \tag{4-45}$$

where $\beta \in \{0, 1\}$ is the criteria coefficient and is used for the purpose of single- or multicriteria decision-making. When a criterion is supposed to be counted in the decision-making process the $\beta = 1$, otherwise $\beta = 0$.

4.2.2.7   Calculating the GIG

The grey incidence grade is used for final decision-making in the GIA and is calculated as follows:

$$GIG_i = \frac{1}{1 + (D_i^\downarrow / D_i^\uparrow)^2}, \quad i = 1, 2, ..., m \tag{4-46}$$

It is obvious that $0 \leq GIG_i \leq 1$. Therefore as GIG_i approaches 1, we are approaching the ideal solution; on the contrary if GIG_i approaches zero, we are getting closer to the anti-ideal solution.

4.3 Case Studies

Choose the best type of prime mover for a CCHP system among the four options of conventional separate production (CSP), gas-fired internal combustion engine (IC), MGT, and Stirling engine (STR) for a four-floor, eight-unit residential building with a total living area of 1200 m^2 [10]. The building is assumed to be located in one of five cities (Kerman, Ahwaz, Bandar Anzali, Chabahar, Kamyaran) with five different climates. The weather information including the maximum and minimum dry-bulb temperature (T_{db}) and relative humidity (RH) of the selected cities are given in Table 4.4. In addition the annual average aggregated thermal demand (ATD), electrical demand (E_{dem}), and power to heat ratio (PHR) of the building for the five case studies are given in Figure 4.5.

4.3.1 Solution

In order to find the best type of prime mover, we apply both the fuzzy-MCDM and grey-MCDM methods to the problem, to be sure about the decision. If both methods recommend the same solution it confirms the decision. To do the calculations, MATLAB code was generated for the two MCDM methods, but here in this book some hand calculations are also presented as a guideline.

Table 4.4 Weather Information for the Representative Cities

City	$T_{db\_max}(^{o}C)$	$T_{db\_min}(^{o}C)$	$RH_{max}(\%)$	$RH_{min}(\%)$
Kerman	36.51	-2.50	78	10
Ahwaz	46.74	8.73	86.13	8.76
Bandar Anzali	29.76	4.61	96.2	64.36
Chabahar	33.55	16.87	86.10	43.14
Kamyaran	36.82	-4.88	85.05	11.38

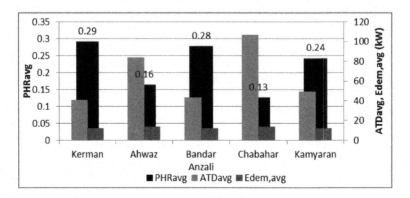

Figure 4.5 The annual average PHR, ATD, and E_{dem} for the five case studies.

Table 4.5 The Judgment Matrix J for the Fuzzy-MCDM and Grey-MCDM [6, 7, 11-24]

Main Criteria	Subcriteria	Fuzzy-MCDM				Grey-MCDM			
		CSP	*IC*	*MGT*	*STR*	*CSP*	*IC*	*MGT*	*STR*
Technological	$\eta_{o\text{-PM}}$	0.52*	0.9	0.85	0.95	0.52	0.9	0.85	0.95
	PHR	Δ**	0.4	0.55	0.3	Δ**	0.4	0.55	0.3
	Maturity	VH	VH	M	L	A	A	C	D
	OPL	M	H	M	H	C	B	C	B
	UFCR	H	H	M	M	B	B	C	C
Economic	I	L	M	H	VH	D	C	B	A
	I_{OM}	VL	H	M	L	E	B	C	D
	NPV	L***	VH	L	M	D	A	D	C
Environmental	NO_x	300	0.7	0.15	0.23	300	0.7	0.15	0.23
	CO_2	700	360	740	250	700	360	740	250
	CO	4	0.8	0.54	0.42	4	0.8	0.54	0.42
	Noise	VL	H	M	L	E	B	C	D
Miscellaneous	IEL	L	L	VH	H	D	D	A	B
	Footprint	0.06	0.05	0.03	0.03	0.06	0.05	0.03	0.03
	Lifetime	10	20	12	10	10	20	12	10
	LAEM	H	VH	VL	M	B	A	E	C

*If a fuzzy number is shown by a single real number, it means that the three components of the fuzzy number are equal, for example 0.52 = (0.52, 0.52, 0.52).
**This is the PHR of the building in different climates that is presented in Figure 4.5
***The above table is prepared based on the user's profit.

As the first step, the criteria and subcriteria that are involved in the decision should be introduced. For this purpose the judgment matrix for the fuzzy-MCDM and grey-MCDM are presented in Table 4.5.

The economic data that is used in the decision-making process is based on a qualitative evaluation that makes use of manufacturers' data and literature review. It should be understood that in the decision-making stage, very precise data is not necessary, because we are comparing some technologies from different points of view. This is also an advantage of the MCDM methods. By using data from the literature, we are able to compare these technologies. For example, Table 2 of Ref. [11] presents some quantitative data for the economic parameters such as average I and I_{OM}. In addition, Table 3 of Ref. [6] presents some quantitative data for I, payback period, NPV, and total annual cost. Furthermore, Table 2 of Ref [7] presents some quantitative data for I, I_{OM}, and service life. By considering these data and considering the impact of import-export limitations on the economic characteristics of some high-tech products such as micro-turbines, a qualitative evaluation is conducted for the economic analysis. The ability of a prime mover to produce recovered heat is also compared according to the PHR in the analysis.

The next step is to normalize the judgment matrix; the normalized matrixes for fuzzy- and grey-MCDM are provided according to the quantification and normalization techniques presented for each method. In the following two examples normalization for both methods is presented.

Example 4.1

Normalize the maturity quality of the MGT in the judgment matrix (Table 4.5) based on fuzzy and grey calculations:

Answer

Because maturity is a benefit, the maximum quality $(J_j^\uparrow)$ among the alternatives should be found. The maximum quality is VH, which equals the fuzzy number $J_3^\uparrow = (0.7, 1.0, 1.0)$. The maturity quality of MGT is M, which corresponds to $J_{23} = (0.3, 0.5, 0.7)$. Therefore

$$J_{23}^N = \frac{J_{23}}{J_3^\uparrow} = \left(\frac{0.3}{1.0}, \frac{0.5}{1.0}, \frac{0.7}{0.7}\right) = (0.3, 0.5, 1)$$
$$\therefore \quad J_{23}^N = (0.3, 0.5, 1)$$

For the grey-MCDM normalization, the maximum and minimum qualities are required. The maximum and minimum qualities of maturity are A = 0.9 and D = 0.3, respectively. The maturity quality of MGT is C = 0.5. Therefore

$$J_{23}^N = \frac{J_{23} - J_3^\downarrow}{J_3^\uparrow - J_3^\downarrow} = \frac{0.5 - 0.3}{0.9 - 0.3} = 0.33$$
$$\therefore \quad J_{23}^N = 0.33$$

The normalized weight matrixes for the fuzzy- and grey-MCDM are presented in Tables 4.6A and B.

Table 4.6A **Normalized Judgment Matrix for the Fuzzy-MCDM**

Subcriteria	Fuzzy-MCDM			
	CSP	IC	MGT	STR
η_0-PM	(0.55, 0.55, 0.55)	(0.95, 0.95, 0.95)	(0.89, 0.89, 0.89)	(1, 1, 1)
PHR	(1, 1, 1)	Δ /0.4	Δ /0.55	Δ /0.3
Maturity	(0.7, 1, 1)	(0.7, 1, 1)	(0.3, 0.5, 1)	(0, 0.3, 0.71)
OPL	(0.3, 0.71, 1)	(0.5, 1, 1)	(0.3, 0.71, 1)	(0.5, 1, 1)
UFCR	(0.5, 1, 1)	(0.5, 1, 1)	(0.3, 0.71, 1)	(0.3, 0.71, 1)
I	(0, 1, 1)	(0, 0.6, 1)	(0, 0.43, 1)	(0, 0.3, 0.71)
I_{OM}	(0, 1, 1)	(0, 0, 0.6)	(0, 0, 1)	(0, 0, 1)
NPV	(0, 0.3, 0.71)	(0.7, 1, 1)	(0, 0.3, 0.71)	(0.3, 0.5, 1)
NO_x	(0.05, 0.05, 0.05)	(0.21, 0.21, 0.21)	(1, 1, 1)	(0.65, 0.65, 0.65)
CO_2	(0.36, 0.36, 0.36)	(0.69, 0.69, 0.69)	(0.34, 0.34, 0.34)	(1, 1, 1)
CO	(0.1, 0.1, 0.1)	(0.52, 0.52, 0.52)	(0.78, 0.78, 0.78)	(1, 1, 1)
Noise	(0, 0, 0.3)	(0, 0, 0.6)	(0, 0, 1)	(0, 0, 1)
IEL	(0, 1, 1)	(0, 1, 1)	(0, 0.3, 0.71)	(0, 0.43, 1)
Footprint	(0.5, 0.5, 0.5)	(0.6, 0.6, 0.6)	(1, 1, 1)	(1, 1, 1)
Lifetime	(0.5, 0.5, 0.5)	(1, 1, 1)	(0.6, 0.6, 0.6)	(0.5, 0.5, 0.5)
LAEM	(0.5, 0.7, 1)	(0.7, 1, 1)	(0, 0, 0.43)	(0.3, 0.5, 1)

Table 4.6B Normalized Judgment Matrix for the Grey-MCDM

Subcriteria	Grey-MCDM			
	CSP	IC	MGT	STR
ηo_{-PM}	0	0.88	0.76	1
PHR	1	Δ /0.4	Δ /0.55	Δ /0.3
Maturity	1	1	0.33	0
OPL	0	1	0	1
UFCR	1	1	0	0
I	1	0.67	0.33	0
I_{OM}	1	0	0.34	0.67
NPV	0	1	0	0.33
NO_x	0	0.99	1	0.99
CO_2	0.08	0.77	0	1
CO	0	0.89	0.96	1
Noise	1	0	0.33	0.67
IEL	1	1	0	0.33
Footprint	0	0.33	1	1
Lifetime	0	1	0.2	0
LAEM	0.75	1	0	0.5

The third step for the decision is to calculate the normalized weight matrix. For this purpose the pairwise comparison matrix should be provided first as in Tables 4.7A through E. In Table 4.7A, for example, the decision-maker's opinion is that the economic criterion is strongly more important (SMI) than the environmental criterion; therefore these tables present the degree of importance of the columns with respect to the rows. Different decision-makers may have different opinions about this table, because the opinion depends on many criteria, such as the country that the CCHP is going to be installed in or designed for.

Table 4.7A Pairwise Comparison of the Main Criteria

	Technological	Economic	Environmental	Miscellaneous
Technological	JE	WMI	SMI(R)	EI
Economic	WMI(R)	JE	SMI(R)	EI
Environmental	SMI	SMI	JE	SMI
Miscellaneous	EI(R)	EI(R)	SMI(R)	JE

Table 4.7B Pairwise Comparison of the Technological Subcriteria

	η_{o-PM}	PHR	Maturity	OPL	UFCR
η_{o-PM}	JE	VSMI(R)	WMI(R)	WMI(R)	VSMI(R)
PHR	VSMI	JE	SMI	WMI	WMI(R)
Maturity	WMI	SMI(R)	JE	SMI(R)	SMI(R)
OPL	WMI	WMI(R)	SMI	JE	WMI(R)
UFCR	VSMI	WMI	SMI	WMI	JE

Table 4.7C Pairwise Comparison of the Economic Subcriteria

	I	I_{OM}	NPV
I	JE	EI	VSMI
I_{OM}	EI(R)	JE	VSMI
NPV	VSMI (R)	VSMI (R)	JE

Table 4.7D Pairwise Comparison of the Environmental Subcriteria

	NO_x	CO_2	CO	Noise
NO_x	JE	EI	EI	SMI(R)
CO_2	EI(R)	JE	EI	SMI(R)
CO	EI(R)	EI(R)	JE	SMI(R)
Noise	SMI	SMI	SMI	JE

Table 4.7E Pairwise Comparison of the Miscellaneous Subcriteria

	IEL	Footprint	Lifetime	LAEM
IEL	JE	AMI(R)	AMI(R)	AMI(R)
Footprint	AMI	JE	VSMI	VSMI
Lifetime	AMI	VSMI(R)	JE	EI
LAEM	AMI	VSMI(R)	EI(R)	JE

Example 4.2

Calculate the normalized weight for the initial investment cost (I) according to the fuzzy-MCDM

Answer

Because I is an economic criterion, the *PC* matrix of the economic criterion should be quantified as in Table 4.8.

Then the eigenvectors of the *PC* matrix are given as follows:

$$E_1 = \left[(1,1,1)\times(2/3,1,2)\times(1/3,2/5,1/2)\right]^{1/3} = \left[(2/9,2/5,1)\right]^{1/3}$$

$$= (0.6057, \quad 0.7368, \quad 1.0000)$$

$$\therefore E_1 = (0.6057, \quad 0.7368, \quad 1.0000)$$

Table 4.8 Quantified Pairwise Comparison of the Economic Subcriteria

	I	I_{OM}	NPV
I	(1,1,1)	(1/2,1,3/2)	(2,5/2,3)
I_{OM}	(2/3,1,2)	(1,1,1)	(2,5/2,3)
NPV	(1/3,2/5,1/2)	(1/3,2/5,1/2)	(1,1,1)

$$E_2 = \left[(1/2, 1, 3/2) \times (1, 1, 1) \times (1/3, 2/5, 1/2)\right]^{1/3} = \left[(1/6, 2/5, 3/4)\right]^{1/3}$$
$$= (0.5503 \quad 0.7368 \quad 0.9086)$$
$$\therefore E_2 = (0.5503 \quad 0.7368 \quad 0.9086)$$

$$E_3 = \left[(2, 5/2, 3) \times (2, 5/2, 3) \times (1, 1, 1)\right]^{1/3} = \left[(4, 25/4, 9)\right]^{1/3}$$
$$= (1.5874 \quad 1.8420 \quad 2.0801)$$
$$\therefore E_3 = (1.5874 \quad 1.8420 \quad 2.0801)$$

And finally the weight of I would be

$$W_1 = \frac{E_1}{\sum_{i=1}^{3} E_i}$$

$$= \frac{(0.6057, 0.7368, 1.0000)}{(0.6057, 0.7368, 1.0000) + (0.5503, 0.7368, 0.9086) + (1.5874, 1.8420, 2.0801)}$$

$$= \frac{(0.6057, 0.7368, 1.0000)}{(2.7432, 3.3156, 3.9887)} = (0.1518, 0.2222, 0.3645)$$

$$\therefore \quad W_1 = (0.1518, 0.2222, 0.3645)$$

This is the fuzzy weight that is presented in Figure 4.7.

Approximately the same procedure can be applied in the grey-MCDM to calculate the normalized weight matrix elements.

The normalized weights of all criteria and subcriteria in fuzzy- and grey- MCDM are given in Figures 4.6 and 4.7 and Table 4.9. It should be noted here that the fuzzy weighting only feeds from the decision-maker's opinion, therefore there is no difference between the weights in the five climates. However, the weighting method employed in the grey-MCDM feeds from the decision-maker's opinion and judgment matrix as well, therefore due to $PHR_{building}$, which is different in the five climates, there are different weights proposed by EIM and the combined weighting methods for the five cities.

Continuing the calculations described in Examples 4.1 and 4.2 the GIG and CN, which are the decision-making parameters of the grey- and fuzzy-MCDM, are calculated. The results are presented in Table 4.10 for both methods.

As the results show, both fuzzy- and grey-MCDM propose the internal combustion engine as the best prime mover for the five climates. But paying attention to the single main criterion results, they reveal interesting facts. From the technological point of view, the methods recommend using IC, but they propose different prime

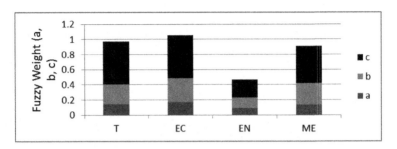

Figure 4.6 Normalized fuzzy weight components of the main criteria.

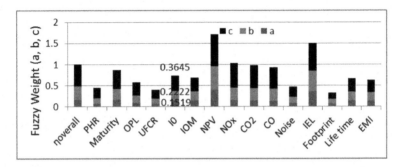

Figure 4.7 Normalized fuzzy weight components of the subcriteria.

movers when talking about the economic criterion; the grey-MCDM proposes CSP while the fuzzy-MCDM recommends using IC. Based on the environmental criterion, both methods recommend using a Stirling engine. The IC is recommended by the miscellaneous criterion in the two methods. Finally the integrated or multicriteria evaluations recommend using the IC for all five climates in Iran. It must be noted that this decision is only valid for these cities. For other countries some subcriteria may be omitted and some new subcriteria be added, the magnitude or quality of some subcriteria may change, and consequently the GIG and CN may change.

4.4   The Effect of μ and ξ on λ and GIG

In the cases studied above it is assumed that $\mu = 0.5$ and $\xi = 0.5$, but as stated these coefficients fall between 0 and 1. For this reason, in this section, the impact of changing these parameters on λ and the GIG is studied. The effect of μ on λ and the GIG is presented in Figures 4.8 and 4.9, respectively. As can be seen, when μ approaches zero, the contribution of the AHP (λ_1) and the EIM (λ_2) in the combined weighting method is the same but when μ approaches 1, the EIM contribution is more dominant. Figure 4.9 shows that when μ changes from 0 to 0.99, the GIG for the four alternatives experiences slight changes. However, the results show that the priority of the prime mover proposed by the GIA does not change although the GIG changes. Figure 4.10

Table 4.9 Normalized Weight of All Criteria and Subcriteria According to the Weighting Technique of the Grey-MCDM for the Five Cities

→Main Criteria		Technological: 0.243					Economic: 0.308			Environmental: 0.175				Miscellaneous: 0.274			
↓Cities	→Subcriteria	$\eta_{overall}$	PHR	Maturity	OPL	UFCR	I_0	I_{OM}	NPV	NO_x	CO_2	CO	Noise	IEL	Footprint	Life time	EMI
Kerman	AHP	0.313	0.139	0.260	0.178	0.110	0.221	0.257	0.523	0.248	0.286	0.329	0.137	0.508	0.088	0.185	0.219
	EIM	0.138	0.204	0.496	0.081	0.081	0.202	0.486	0.311	0.657	0.043	0.216	0.084	0.293	0.121	0.121	0.465
	Combined	0.203	0.180	0.409	0.117	0.092	0.209	0.402	0.389	0.506	0.133	0.257	0.104	0.372	0.109	0.145	0.374
Ahwaz	AHP	0.313	0.139	0.260	0.178	0.110	0.221	0.257	0.523	0.248	0.286	0.329	0.137	0.508	0.088	0.185	0.219
	EIM	0.108	0.377	0.389	0.063	0.063	0.202	0.486	0.311	0.657	0.043	0.216	0.084	0.293	0.121	0.121	0.465
	Combined	0.188	0.284	0.339	0.108	0.081	0.209	0.397	0.393	0.498	0.137	0.260	0.105	0.376	0.108	0.146	0.370
Bandar Anzali	AHP	0.313	0.139	0.260	0.178	0.110	0.221	0.257	0.523	0.248	0.286	0.329	0.137	0.508	0.088	0.185	0.219
	EIM	0.136	0.217	0.488	0.080	0.080	0.202	0.486	0.311	0.657	0.043	0.216	0.084	0.293	0.121	0.121	0.465
	Combined	0.201	0.188	0.404	0.116	0.091	0.209	0.402	0.389	0.506	0.133	0.257	0.104	0.372	0.109	0.145	0.374
Chabahar	AHP	0.313	0.139	0.260	0.178	0.110	0.221	0.257	0.523	0.248	0.286	0.329	0.137	0.508	0.088	0.185	0.219
	EIM	0.097	0.444	0.347	0.057	0.057	0.202	0.486	0.311	0.657	0.043	0.216	0.084	0.293	0.121	0.121	0.465
	Combined	0.183	0.322	0.312	0.105	0.078	0.210	0.395	0.396	0.494	0.140	0.261	0.105	0.379	0.108	0.147	0.367
Kamyaran	AHP	0.313	0.139	0.260	0.178	0.110	0.221	0.257	0.523	0.248	0.286	0.329	0.137	0.508	0.088	0.185	0.219
	EIM	0.129	0.259	0.462	0.075	0.075	0.202	0.486	0.311	0.657	0.043	0.216	0.084	0.293	0.121	0.121	0.465
	Combined	0.197	0.214	0.387	0.113	0.088	0.209	0.401	0.390	0.505	0.134	0.258	0.104	0.373	0.109	0.145	0.373

Table 4.10 Single- and Multicriteria Decision-Making Results Regarding Choice of the Best Prime Mover According to the Fuzzy-MCDM and Grey-MCDM for the Five Cities

Five Cities

GIG (Grey-MCDM)

Criteria →	Kerman				Ahwaz				Chabahar				Kamyaran				Bandar Anzali			
	CSP	IC	MGT	STR	CSP	IC	MGT	STR	CSP	IC	MGT	STR	CSP	IC	MGT	STR	CSP	IC	MGT	STR
T	0.675	0.854	0.279	0.489	0.697	0.800	0.248	0.468	0.703	0.777	0.238	0.468	0.683	0.837	0.268	0.472	0.677	0.850	0.276	0.482
EC	0.610	0.522	0.212	0.376	0.606	0.527	0.210	0.375	0.603	0.529	0.210	0.374	0.609	0.523	0.211	0.376	0.609	0.522	0.211	0.376
EN	0.165	0.801	0.733	0.886	0.166	0.799	0.728	0.886	0.166	0.798	0.725	0.886	0.165	0.800	0.732	0.886	0.165	0.801	0.733	0.886
S	0.642	0.870	0.179	0.420	0.643	0.870	0.178	0.419	0.643	0.870	0.178	0.418	0.642	0.870	0.179	0.420	0.642	0.870	0.179	0.420
MCDM	0.512	0.778	0.339	0.577	0.517	0.763	0.328	0.572	0.519	0.757	0.324	0.572	0.514	0.773	0.336	0.573	0.512	0.777	0.338	0.576

CN (Fuzzy-MCDM)

Criteria →	Kerman				Ahwaz				Chabahar				Kamyaran				Bandar Anzali			
	CSP	IC	MGT	STR	CSP	IC	MGT	STR	CSP	IC	MGT	STR	CSP	IC	MGT	STR	CSP	IC	MGT	STR
T	0.427	1.000	0.000	0.367	0.601	1.000	0.000	0.264	0.656	1.00	0.000	0.230	0.492	1.000	0.000	0.329	0.445	1.000	0.000	0.357
EC	0.664	1.000	0.000	0.287	0.664	1.000	0.000	0.287	0.664	1.00	0.000	0.287	0.664	1.000	0.000	0.287	0.664	1.000	0.000	0.287
EN	0.000	0.425	0.780	1.000	0.000	0.425	0.780	1.000	0.000	0.425	0.780	1.000	0.000	0.425	0.780	1.000	0.000	0.425	0.780	1.000
S	0.648	1.00	0.000	0.315	0.648	1.00	0.000	0.315	0.648	1.00	0.000	0.315	0.648	1.00	0.000	0.315	0.648	1.00	0.000	0.315
MCDM	0.103	1.00	0.046	0.652	0.135	1.00	0.025	0.621	0.145	1.00	0.018	0.611	0.115	1.00	0.038	0.640	0.106	1.00	0.043	0.649

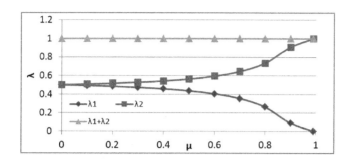

Figure 4.8 Impact of variation of μ on λ_1 and λ_2.

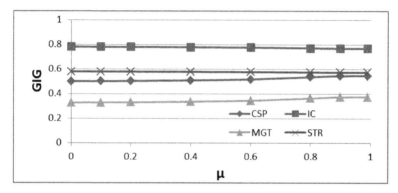

Figure 4.9 Impact of variation of μ on the GIG for $\xi = 0.5$.

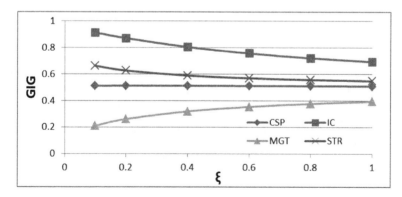

Figure 4.10 Impact of variation of ξ on the GIG for $\mu = 0.5$.

shows the impact of the distinguishing coefficient on the GIG. It shows that when the ξ approaches 1 the GIG for different prime movers approaches 0.5, but the priority for the prime movers recommended by the GIA remains unchanged. Therefore, as a result changing the μ and ξ does not change the decision, but changes the magnitude of the GIG. When the μ and ξ approach 1, the GIG approaches 0.5.

4.5 Problems

1. Repeat the above case study for the city you live in.
2. Repeat Problem 1 after adding a fuel cell as the fifth prime mover type.
3. Choose the best type of cooling system among indirect fired, single-, double-, and triple-absorption chillers, and adsorption chiller for the prime mover type selected in the previous problem. What criteria are involved? Does the solution and refrigerant type make any difference in the decision? What about the quality of the recoverable heat from the prime mover?
4. Does the heat recovery system differ for an internal combustion engine and a MGT?

References

[1] Liu, Sifeng, Lin, Yi, 2011. Grey Systems: Theory and Applications. Springer. Berlin Heidelberg, ISBN 978-3-642-16158-2.
[2] Zadeh, L.A., 1965. Fuzzy Sets. Information and Control 3, 338–353.
[3] Help product from software MATLAB R2008a..
[4] Kahraman, C., Ertay, T., Buyukozkan, G., 2006. A Fuzzy Optimization Model for QFD Planning Process Using Analytic Network Approach. European Journal of Operational Research 171, 390–411.
[5] Ebrahimi, M., Keshavarz, A., 2012. Prime Mover Selection for a Residential Micro-CCHP by Using Two Multi-Criteria Decision-Making Methods. Energy and Buildings 55, 322–331.
[6] Jing, Y.-Y., Bai, H., Wang, J.-J., 2012. A Fuzzy Multi-Criteria Decision-Making Model for CCHP Systems Driven by Different Energy Sources. Energy Policy 42, 286–296.
[7] Wang, J.-J., Jing, Y.-Y., Zhang, C.-F., Zhang, X.-T., Shi, G.-H., 2008. Integrated Evaluation of Distributed Triple-Generation Systems Using Improved Grey Incidence Approach. Energy 33, 1427–1437.
[8] Y. Lin, SF Liu, "A Historical Introduction to Grey Systems Theory," *Proceedings of the IEEE International Conference on Systems, Man and Cybernetics*, Hague, the Netherlands, vol. 3. New York: Institute of Electrical and Electronics Engineers Inc. 2004, p. 2403–2408.
[9] Liu, S.F., Lin, Y., 1998. An Introduction to Grey Systems Theory. IIGSS Academic Publisher, Grove City.
[10] Ebrahimi, M., Keshavarz, A., 2012. Climate Impact on the Prime Mover Size and Design of a CCHP System for the Residential Building. Energy and Buildings 54, 283–289.
[11] Wu, D.W., Wang, R.Z., 2006. Combined Cooling, Heating, and Power: A Review. Progress in Energy and Combustion Science 32, 459–495.
[12] Kong, X.Q., Wang, R.Z., Huang, X.H., 2004. Energy Efficiency and Economic feasibility of CCHP Driven by Stirling Engine. Energy Conversion and Management 45, 1433–1442.
[13] Katsigiannis, P.A., Papadopoulos, D.P., 2005. A General Technoeconomic and Environmental Procedure for Assessment of Small-Scale Cogeneration Scheme Installations: Application to a Local Industry Operating in Thrace, Greece, Using Microturbines. Energy Conversion and Management 46, 3150–3174.
[14] Onovwiona, H.I., Ugursal, V.I., 2006. Residential Cogeneration Systems: Review of the Current Technology. Renewable and Sustainable Energy Reviews 10, 389–431.
[15] Yagoub, W., Doherty, P., Riffat, S.B., 2006. Solar Energy-Gas Driven Micro-CHP System for an Office Building. Applied Thermal Engineering 26, 1604–1610.
[16] Godefroy, J., Boukhanouf, R., Riffat, S., 2007. Design, Testing and Mathematical Modeling of a Small-Scale CHP and Cooling System (Small CHP-Ejector Trigeneration). Applied Thermal Engineering 27, 68–77.

[17] Huangfu, Y., Wu, J.Y., Wang, R.Z., Kong, X.Q., Wei, B.H., 2007. Evaluation and Analysis of Novel Micro-Scale Combined Cooling, Heating and Power (MCCHP) System. Energy Conversion and Management 48, 1703–1709.

[18] Pehnt, M., 2008. Environmental Impacts of Distributed Energy Systems—The Case of Micro Cogeneration. Environmental Science & Policy 11, 25–37.

[19] Kuhn, V., Klemes, J., Bulatov, I., 2008. MicroCHP: Overview of Selected Technologies, Products and Field Test Results. Applied Thermal Engineering 28, 2039–2048.

[20] Sugiartha, N., Tassou, S.A., Chaer, I., Marriott, D., 2009. Trigeneration in Food Retail: An Energetic, Economic and Environmental Evaluation for a Supermarket Application. Applied Thermal Engineering 29, 2624–2632.

[21] Sanaye, S., Ardali, M.R., 2009. Estimating the Power and Number of Microturbines in Small-Scale Combined Heat and Power Systems. Applied Energy 86, 895–903.

[22] Ren, H., Zhou, W., Nakagami, K., Gao, W., 2010. Integrated Design and Evaluation of Biomass Energy System Taking into Consideration Demand Side Characteristics. Energy 35, 2210–2222.

[23] Tichi, S.G., Ardehali, M.M., Nazari, M.E., 2010. Examination of Energy Price Policies in Iran for Optimal Configuration of CHP and CCHP Systems Based on Particle Swarm Optimization Algorithm. Energy Policy 38, 6240–6250.

[24] Monteiro, E., Afonso Moreira, N., Ferreira, S., 2009. Planning of Micro-Combined Heat and Power Systems in the Portuguese Scenario. Applied Energy 86, 290–298.

CCHP Load Calculations

5.1   Introduction

After making a decision about the component types for the CCHP system, component sizing is the next design step. Apart from the choice of sizing method, the consumer demands or loads play the most important role in the sizing of components. Therefore, determining consumer demands is an essential step in designing a CCHP system. Different sizing methods may use different load information, and they can be classified as follows:

- Single number method (SNM): Some methods only use yearly peak loads of electricity, cooling, heating, and DHW demands. Therefore, in this case only one number is available for each load type as the input data for the sizing method.
- 288 number method (TNM): Some methods may use a 24-hour average load calculation for each month. In these methods 12 sets of 24-hour data ($12 \times 24 = 288$ numbers for each load type) will be available as input data for sizing the CCHP components.
- 8760 number method (ENM): Some methods may use hourly load calculations during the whole year. Therefore, in these methods $365 \times 24 = 8760$ numbers are available for each load type as input data for the sizing method.

SNM is simple to calculate and can be considered a rough estimation method for sizing and initial capital costs. This method does not give detailed information about fuel consumption, cost, overall efficiency, and environmental benefits. Moreover, since in these methods every component is designed according to the peak loads, the components may be oversized.

TNM is more accurate than SNM. But this method does not trace load changes during a month because it gives only 24 numbers corresponding to 24 hours for each month (in other words it considers all days of each month the same). The accuracy of results in this method suffers from lack of daily information during month.

ENM is the most currently accurate available load calculation method, because it calculates the loads for every hour during a year. Calculation of fuel consumption, fuel savings, economic evaluations, etc., is much more accurate than SNM and TNM. The impact of load variation on the economic, thermodynamic, and environmental criteria can be investigated precisely.

5.2   Weather Information

In order to calculate the electricity, heating, cooling, and DHW loads of a consumer, it is important to know the building orientation, construction materials of the building and their characteristics, altitude, latitude, longitude, number of occupants, presence time and schedule, activity type, lighting type, lighting index per square meter, electricity consumers, and corresponding heat generation. In addition, having the daily/hourly average value of the last 5 to 10 years of dry bulb temperature and relative humidity of the ambient where the consumer is located is essential. The wet bulb temperature, which some load calculators utilize for load estimations, can be calculated according to the dry bulb temperature and relative humidity.

The weather information of different locations is usually available on national metrological organization websites. In addition to the national websites some international websites exist that give weather information in most of locations around the world. Some of these websites are mentioned here:

http://www.accuweather.com
http://www.wunderground.com
http://www.intellicast.com
http://www.weather-forecast.com
http://www.weather.gov
http://www.xcweather.co.uk
http://www.metoffice.gov.uk
http://forecast.io
http://www.yr.no
http://www.weather-forecast.com
http://weather.weatherbug.com

The weather information and building characteristics can be used in the load calculator software to calculate the building demands.

A list of software that can be used for load calculations is introduced in the next section [1].

5.3   Load Calculators

The following software is used for load calculations in determining the electiricty, cooling, heating, and DHW demands of a consumers:

AIRWIND Pro
BTU Analysis Plus
BTU Analysis REG
BV2
CAMEL
CBE UFAD Cooling Design Tool
CHVAC
CL4M Commercial Cooling and Heating Loads
Climawin 2005
CoDyBa
Cold Room Calc
COLDWIND Pro
COMFIE
Cool Room Calc

Curb
Cymap Mechanical
DesignBuilder
DeST
DPClima
E-Z Heatloss
ECOTECT
ENER-WIN
Energy Profiler
Energy Profiler Online
Energy Scheming
Energy Usage Forecasts
EnergyPlus
EnergySavvy
ESP-r
HAP
HAP System Design Load
HBLC
HEED
HOT2000
HVAC Residential Load Calcs – HD for the iPad
ISOVER Energi
J-Works
LESO-COMFORT
LESOCOOL
LESOKAI
LESOSAI
Load Express
Micropas6
PASSPORT
QwickLoad
RHVAC
Right-Suite Residential for Windows
RIUSKA
System Analyzer
Toolkit for Building Load Calculations
TOP Energy
TRACE Load 700

In the following a case study including five climates is presented to show the load calculation steps.

5.4   Load Calculation Example

In this book, in order to investigate the impact of climate on load calculations and CCHP design, an identical building is considered in five different climates presented in Table 5.1.

Table 5.1 **Five Different Climates and Nominated Cities**

Climate	Nominated City
Tropical and dry in summer, cold in winter (TDC)	Kerman
Temperate and dry in summer, extremely cold in winter (TDEC)	Kamyaran
Temperate and humid in summer, cold in winter (THC)	Bandar-Anzali
Tropical and humid in summer, temperate in winter (THT)	Chabahar
Tropical and semi-humid in summer, temperate in winter (TSHT)	Ahwaz

It is important to mention that considering nominated cities from Iran is just due to the authors' familiarity with these climates and availability of weather information. This method of decision-making, design, and optimization is completely general and can be applied to any other city or climates around the world. To start the analyses, the daily weather information from the beginning of August 2006 to the end of July 2011 of the five cities is gathered from the available archive at the Iran Metrological Organization website [2]. The information includes the daily maximum and minimum of dry bulb temperature (T_{db}) and relative humidity (RH). The monthly average of the T_{db} and RH for the five years is calculated, and finally the average of $T_{db,min}$, $T_{db,max}$, and RH of each specific month during the five years is determined. The five-year averages of T_{db} and RH at the atmospheric pressure are used in the engineering equation solver (EES) to determine the corresponding wet bulb temperature (T_{wb}). The weather information for Kerman is presented in Figures 5.1 to 5.5 as examples but the results of other cities are presented in appendix 1 in Figures A1.1 to A1.12.

In the simulations presented in the next chapters, knowing the hourly temperature is necessary. The hygrometric data reported above are curve fitted using the MATLAB curve-fitting tool for each climate. The equations for $T_{db,min}$ and $T_{db,max}$ for the five climates are

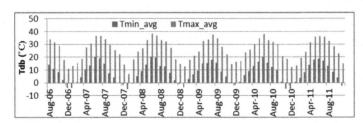

Figure 5.1 Dry bulb temperature over five years for Kerman.

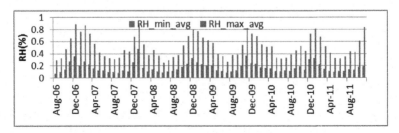

Figure 5.2 Relative humidity over five years for Kerman.

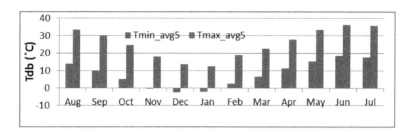

Figure 5.3 Five-year average for T_{DB} for Kerman.

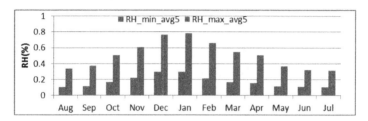

Figure 5.4 Five-year average for RH for Kerman.

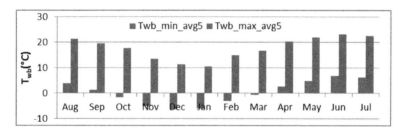

Figure 5.5 T_{wb} of Kerman based on five-year average of RH and T_{db}.

given in Eqs. (5-1) to (5-5). In these equations N is the day number starting from N = 1 for January 1st to N = 365 at the end of year. In order to calculate the hourly temperature for each day Erbs's model [3] is utilized (Eq. 5-6). This equation calculates the hourly temperature during each day according to the $T_{db,min}$ and $T_{db,max}$ of the corresponding day.

In the mathematical simulations for every climate and city such curve fittings can be done and the equation can be used in the simulation.

Kerman

$$T_{db,min}(N) = q_1N^6 + q_2N^5 + q_3N^4 + q_4N^3 + q_5N^2 + q_6N + q_7$$
$$T_{db,max}(N) = p_1N^6 + p_2N^5 + p_3N^4 + p_4N^3 + p_5N^2 + p_6N + p_7$$
$$q_1 = -7.35 \times 10^{-13}; q_2 = 8.156 \times 10^{-10}; q_3 = -3.241 \times 10^{-7}$$
$$q_4 = 5.464 \times 10^{-5}; q_5 = -41.21 \times 10^{-4}; q_6 = 0.2764$$
$$q_7 = -5.539; p_1 = -5.879 \times 10^{-13}; p_2 = 6.716 \times 10^{-10}$$
$$p_3 = -2.776 \times 10^{-7}; p_4 = 4.931 \times 10^{-5}; p_5 = -41.1 \times 10^{-4}$$
$$p_6 = 32.24 \times 10^{-2}; p_7 = 8.655$$

$$(5\text{-}1)$$

Ahwaz

$$T_{db,min}(N) = p_1 N^5 + p_2 N^4 + p_3 N^3 + p_4 N^2 + p_5 N + p_6$$
$$T_{db\,max}(N) = q_1 N^5 + q_2 N^4 + q_3 N^3 + q_4 N^2 + q_5 N + q_6$$
$$p_1 = 4.826 \times 10^{-11};\ p_2 = -2.193 \times 10^{-8};\ p_3 = -2.807 \times 10^{-6} \tag{5-2}$$
$$p_4 = 15.52 \times 10^{-4};\ p_5 = 20.61 \times 10^{-3};\ p_6 = 6.882$$
$$q_1 = 6.074 \times 10^{-11};\ q_2 = -3.57 \times 10^{-8};\ q_3 = 2.028 \times 10^{-6}$$
$$q_4 = 73.73 \times 10^{-5};\ q_5 = 0.1193;\ q_6 = 15.6$$

Chabahar

$$T_{db,min}(N) = p_1 N^5 + p_2 N^4 + p_3 N^3 + p_4 N^2 + p_5 N + p_6$$
$$T_{db,max}(N) = q_1 N^5 + q_2 N^4 + q_3 N^3 + q_4 N^2 + q_5 N + q_6$$
$$p_1 = -2.741 \times 10^{-11};\ p_2 = 3.242 \times 10^{-8};\ p_3 = -1.346 \times 10^{-5} \tag{5-3}$$
$$p_4 = 19.02 \times 10^{-4};\ p_5 = 62.08 \times 10^{-4};\ p_6 = 16.4;\ q_1 = -7.308 \times 10^{-11}$$
$$q_2 = 6.548 \times 10^{-8};\ q_3 = -2.086 \times 10^{-5};\ q_4 = 24.96 \times 10^{-4}$$
$$q_5 = -28.2 \times 10^{-3};\ q_6 = 24.14$$

Kamyaran

$$T_{db,min}(N) = p_1 N^7 + p_2 N^6 + p_3 N^5 + p_4 N^4 + p_5 N^3 + p_6 N^2 + p_7 N + p_8$$
$$T_{db,max}(N) = q_1 N^7 + q_2 N^6 + q_3 N^5 + q_4 N^4 + q_5 N^3 + q_6 N^2 + q_7 N + q_8$$
$$p_1 = -8.94 \times 10^{-15};\ p_2 = 9.869 \times 10^{-12};\ p_3 = -4.008 \times 10^{-9};\ p_4 = 7.118 \times 10^{-7} \tag{5-4}$$
$$p_5 = -4.811 \times 10^{-5};\ p_6 = -38.42 \times 10^{-5};\ p_7 = 0.2643;\ p_8 = -8.653$$
$$q_1 = 2.13 \times 10^{-15};\ q_2 = -3.826 \times 10^{-12};\ q_3 = 2.757 \times 10^{-9};\ q_4 = -9.812 \times 10^{-7}$$
$$q_5 = 17.53 \times 10^{-5};\ q_6 = -15.19 \times 10^{-3};\ q_7 = 0.7505;\ q_8 = -2.802$$

Bandar e Anzali

$$T_{db,min}(N) = p_1 N^6 + p_2 N^5 + p_3 N^4 + p_4 N^3 + p_5 N^2 + p_6 N + p_7$$
$$T_{db\,max}(N) = q_1 N^6 + q_2 N^5 + q_3 N^4 + q_4 N^3 + q_5 N^2 + q_6 N + q_7$$
$$p_1 = -6.484 \times 10^{-13};\ p_2 = 7.342 \times 10^{-10};\ p_3 = -2.984 \times 10^{-7} \tag{5-5}$$
$$p_4 = 4.996 \times 10^{-5};\ p_5 = -30.45 \times 10^{-4};\ p_6 = 0.137;\ p_7 = 3.112$$
$$q_1 = -9.933 \times 10^{-13};\ q_2 = 1.088 \times 10^{-9};\ q_3 = -4.321 \times 10^{-7}$$
$$q_4 = 7.208 \times 10^{-5};\ q_5 = -44.73 \times 10^{-4};\ q_6 = 0.1569;\ q_7 = 7.419$$

$$T_{amb}(N,HR) = T_{avg} + T_{diff}(0.4632\cos(\Gamma\text{-}3.805) +$$
$$0.0984\cos(2\Gamma\text{-}0.360) + 0.0168\cos(3\Gamma\text{-}0.822) +$$
$$0.0138\cos(4\Gamma\text{-}3.513))$$
$$\Gamma = 2\pi(HR\text{-}1)/24,\ T_{avg} = 0.5\left(T_{db,max}(N) + T_{db,min}(N)\right) \tag{5-6}$$
$$T_{diff} = T_{db,max}(N) - T_{db,min}(N)$$

where HR is the hour from 1 to 24. It is worth mentioning that most of load calculators have the weather information for some sample cities in their database. If the city in which the CCHP is to be designed is included in the database of the load calculator software, there is no need to collect weather information and this step can be skipped.

After determining the weather data, specifying the building's characteristics is essential to calculate the cooling and heating loads of the building.

The cooling and heating loads of a building have direct impact on the size of cooling and heating systems, and indirect impact on the size of the power generation unit. This is due to the heat recovery of energy loss in the prime mover of CCHP systems, which is responsible for providing all or part of the heating or cooling loads. Therefore, paying more attention to the heating and cooling load calculations is highly recommended because they can change the size of every component in CCHP systems.

The authors are aware that buildings should be constructed according to the climate. However, in the case study presented in this text, in order to show the impact of climate difference on the design results of CCHP systems, an identical, hypothetical, four-floor, eight-unit residential building with a total living area of 1200m^2 is considered. The specifications of the building are described in the following.

The average ceiling height and building weight are 2.7 m and 468.7 kg/m^2, respectively. The walls are medium weight with an overall U-value of 1.53 W/(m^2.K). The overall U-values of floors above the unconditioned and conditioned spaces are 0.568 W/(m^2.K) and 2.839 W/(m^2.K), respectively. Each floor has 14 double-glazed windows of the 6 mm argon gap type, and each one has 2×2 (m) area. The total number of occupants is 32 people. The lighting is of the free-hanging fixture type with wattage of 20 W/m^2 and a ballast multiplier of 1.25. The building is simulated in HAP and the cooling and heating loads are determined according to the TNM load calculation (12 sets of 24-hour data).

The magnitude and distribution of electricity consumption have a great impact on the design results for a CCHP system. Therefore the electricity consumers of the building for which the CCHP is designed should be identified, and their electricity consumption should be determined according to the nameplates of the consumers. An approximate consumption schedule listing the electricity consumers also should be considered during weekdays, weekends, holidays, etc.

The main electricity consumers and the corresponding power consumptions are as follows [4-6]. The lighting index is 20 W/m^2, an LCD TV uses 160 W, a washing machine uses 2500 W, a refrigerator uses 130 W, a computer uses 250 W, a cloth iron uses 1000 W, and a vacuum cleaner uses 1000W. A demand factor of 0.76 is considered for the consumers. In addition, the power consumption of the water pump and the chiller's solution pump (sp) and refrigerant pump (rp) are also calculated according to the cooling and heating loads as follows [7]:

$$\dot{W}_{sp} + \dot{W}_{rp} = 0.007 C_{dem,\,max}\,(kW),$$
$$\dot{W}_{water\,pump} = 0.005\max(C_{dem,\,max}, H_{dem,\,max}) \tag{5-7}$$

The home appliances considered above are for a typical residential building. It is evident that for every particular building type (commercial, residential, industrial, educational, etc.) different electricity consumers may be defined. For example the electricity consumers of a residential building are very different with respect to a commercial, official, educational, or industrial building. Therefore for every building type, according to the electricity consumers' type, size, and operation schedule, an electricity consumption distribution can be defined.

In addition to the heating, cooling, and electricity loads, in most building types, especially in the residential sector, preparing domestic hot water (DHW) is of great importance. In order to determine the magnitude and distribution of the DHW load, the hot water consumers, approximate consumption flow rate, and schedule should be determined.

The main DHW consumers and corresponding consumption time are presented in Table 5.2 for the residential building considered in the case study [8].

The building demands, including the cooling (C), heating (H), domestic hot water (D), and electricity (E), are reported in Figures 5.6 to 5.10. These data will be used in the following chapters to design the CCHP system.

As can be seen, the cooling and heating load distribution and magnitude are completely different in the five climates. For example, Chabahar needs cooling during the whole year, while no heating is required. Ahwaz receives the maximum cooling load in summer, and it needs a slight amount of heat in some months. Load distributions for Kamyaran and Kerman are approximately the same, but the peak loads of Kamyaran are larger than Kerman. The cooling loads of Kamyaran and

Table 5.2 DHW Consumers for the Typical Residential Building

Consumer	Flow Rate (liter/min)	Consumption Time in 24 Hours per Person (min) for Spring and Summer	Consumption Time in 24 Hours per Person (min) for Autumn and Winter
Shower	6.3	30	22.5
Bath	1.26	30	22.5
Toilet and hand basin	0.2	20	20
Dish washing Sink in kitchen	0.95	30	30
Sink for foot washing	0.2	10	5

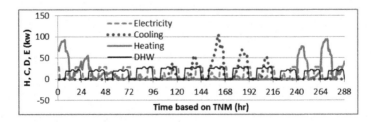

Figure 5.6 Heating, cooling, DHW, and electrical loads for Kerman.

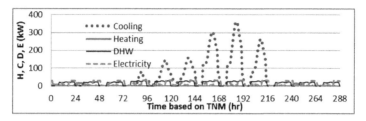

Figure 5.7 Heating, cooling, DHW, and electrical loads for Ahwaz.

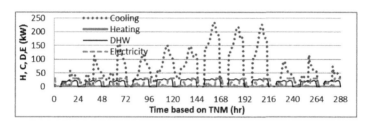

Figure 5.8 Heating, cooling, DHW, and electrical loads for Chabahar.

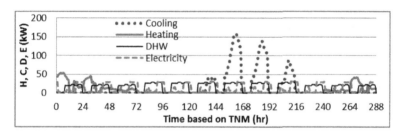

Figure 5.9 Heating, cooling, DHW, and electrical loads for Bandar-Anzali.

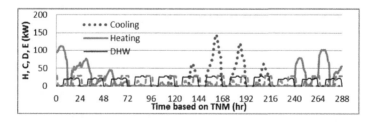

Figure 5.10 Heating, cooling, DHW, and electrical loads for Kamyaran.

Bandar-Anzali are approximately the same, but the heating load of Kamyaran is approximately two times greater than Bandar-Anzali. These differences reveal that the size of CCHP system would be probably different in different climates, and attention must be paid to the load calculations in different climates before designing CCHP systems.

5.5 Problems

1. Provide the last five years of weather data for the city you live in and calculate the average data for the load calculations.
2. Consider a residential or commercial building in your region and calculate the electrical, heating, cooling, and DHW demands for that building.
3. Compare your calculations in Problem 2 with the available energy bills.
4. Curve fit the data you provided in Problem 2.

References

[1] U.S. Department of Energy, Energy Efficiency and Renewable Energy, http://apps1.eere.energy.gov.
[2] Iran Metrological Organization http://www.irimo.ir.
[3] Bilbao, J., De Miguel, A.H., Kambezidis, H.D., 2002. Air Temperature Model Evaluation in the North Mediterranean Belt Area. Journal of Applied Meterology 41, 872–884.
[4] Tehran Regional Energy Distribution Portal, http://www.tvedc.ir.
[5] Energy Department, http://energy.gov.
[6] http://www.wholesalesolar.com.
[7] ASHRAE Handbook– Refrigeration, "Absorption Cooling, Heating, and Refrigeration Equipment," Chapter 41, 2002.
[8] Arthur Bell, Jr., A., 2000. HVAC: Equations, Data, and Rules of Thumb. McGraw-Hill United States of America, ISBN 0-07-136129-4.

CCHP Design

6.1 Introduction

Up to now we have covered two steps of the decision-making about component choice and load calculations in the design of a CCHP system. Sizing is the third step in the design of a CCHP system. The size of a CCHP system has a great impact on its thermodynamic, economic, and environmental characteristics. Since different sizing methods use different criteria, the sizes proposed by different methods are not the same. In this chapter, different sizing methods will be introduced. According to a literature review, methods such as the maximum rectangle method (MRM), developed-MRM, energy management strategy sizing methods, thermodynamic sizing methods, thermoeconomic sizing method,s fitness functions, and the multicriteria sizing methods are used by different researchers and CCHP designers. This diversity in sizing methods creates confusion for CCHP researchers and designers. In the following, the most commonly used sizing methods will be discussed, and in order to compare sizing methods, a case study will be presented using different sizing techniques.

6.2 Maximum Rectangle Method (MRM)

MRM is the most commonly used method for plant choice [1]. The classic MRM is based on the aggregated thermal demand (ATD) versus year hours. In order to utilize MRM, the hourly heating, DHW, and cooling loads of the consumer for which the CCHP is designed must be calculated during a year. Since in CCHP systems all or part of the cooling load is provided by consuming heat in a thermally activated cooling (TAC) system, the heat that is consumed by the TAC system should be considered for calculating the ATD of the building. The ATD is calculated as follows:

$$ATD = H_{dem} + C_{dem}/COP_{TAC} + D_{dem} \tag{6-1}$$

The TAC system may be an absorption/adsorption chiller or dehumidifier.

In order to draw the MRM curve, the ATD values should be sorted from maximum to minimum and then be plotted against the time of year. The result would be as seen in Figure 6.1. In this figure the area under the ATD curve (ATD × Time of year× 8760) is the total supplied heat in kWh to provide the heat demand of the consumer. Some part or all of this heat can be provided by the heat recovery of a prime mover according to the prime mover size and its operation mode (full load or partial load operation).

In Figure 6.1, 100% of time of year corresponds to 8760 hours. Drawing a rectangle with the maximum area under the ATD curve limited to the time and ATD axes

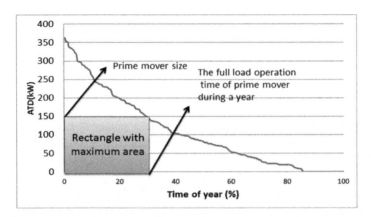

Figure 6.1 Representation of the classic maximum rectangle method.

gives us the recommended prime mover size and its full load operation time during a year (Figure 6.1). The purpose of this method is to supply heat as much as possible to the engine, because it can provide electricity (which can be consumed or sold) and recoverable heat that can be used for cooling or heating purposes. Although some researchers [1] believe that the rectangle area means the amount of heat supplied to the prime mover during a year, in fact it is the kWh_e of electricity production by the prime mover at full load operation.

The area of the rectangle can be calculated according to the following equation:

$$A_{MRM} (kWh_e) = ATD(kW) \times HR$$
$$HR = time \ of \ year(\%) \times 8760(hr)$$

(6-2)

Another curve can be depicted to find the maximum area, as shown in Figure 6.2.

After finding the maximum area according to Figure 6.2, the engine size can be calculated by the following:

$$E_{nom}^{MRM} = \frac{A_{MRM,max}}{HR_{FLO}}$$

(6-3)

In addition A_{MRM} can be plotted against the prime mover size to find the optimum prime mover size recommended by MRM as in Figure 6.3.

The number of hours of full load operation recommended for the prime can also be calculated according to the following:

$$HR_{FLO} \frac{A_{MRM,max}}{E_{nom}^{MRM}}$$

(6-4)

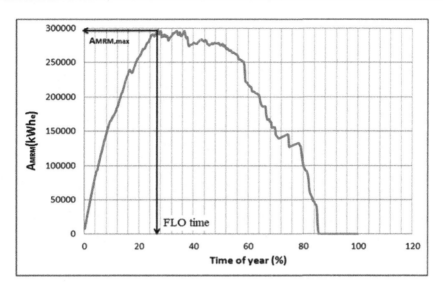

Figure 6.2 Cumulative electricity production (rectangle area) of prime mover versus time of year.

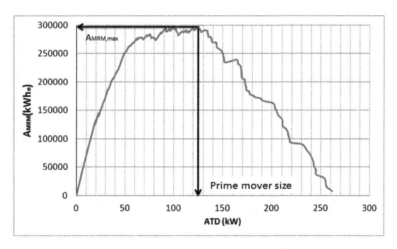

Figure 6.3 Cumulative electricity production (rectangle area) of the prime mover versus its size.

Reference [1] proposes a criteria to increase the full load operation hours until energy saving occurs, or in other words PES ≥ 0 (Figure 6.4). Deriving an equation for this constraint under the following the thermal load (FTL) strategy, discussed in Chapter 1, results in the following equation:

$$\sum_{yearly} F_{CCHP} \leq \sum_{yearly} H_{dem} \bigg/ \left(1 - \frac{\eta_{e,CCHP}}{\eta_{e,pp}.\eta_g} \right) \eta_b \tag{6-5}$$

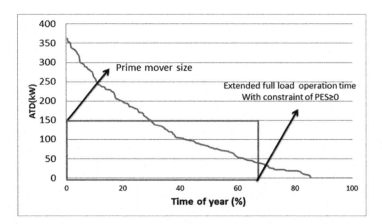

Figure 6.4 Extended time of full load operation under constraints of PES ≥ 0

This means that to satisfy the primary energy saving (PES) constraint under the FTL strategy, Eq. (6-5) should govern.

Considering an energy management strategy such as FTL or following the electricity load (FEL) for the PES constraint while the prime mover size is calculated by MRM seems to be a mistake. The reason is that, for example, when we talk about FTL, it means that the recoverable heat from the engine is sufficient to provide the heating load of the consumer for the whole year. Therefore FTL itself is a sizing method based on the thermal energy management strategy. It is evident that the prime mover size calculated according to MRM is not necessarily equal to the prime mover size that is able to provide all the heating demands during a year (FTL). Therefore the PES constraint proposed by [1] can be modified according to Eq. (6-6) without including any energy management strategy.

$$\sum_{yearly} F_{CCHP} \leq \frac{\sum\limits_{yearly} \zeta}{\left(1 - \dfrac{\eta_{e,CCHP}}{\eta_{e,pp}.\eta_g}\right)\eta_b} \tag{6-6}$$

where

$$\zeta = \begin{cases} H_{dem} & \text{if } Q_{rec} > H_{dem} \\ Q_{rec} & \text{if } Q_{rec} \leq H_{dem} \end{cases} \tag{6-7}$$

In Eq. (6-6) it is assumed that the prime mover efficiency ($\eta_{e,CCHP}$) in full load operation (FLO) and partial load operation (PLO) equals the nominal electrical efficiency.

6.3   Developed-MRM

MRM has been further developed into vertical, horizontal, and high-level analysis [2]. These developments are discussed in more detail in the following subsections.

6.3.1   *Horizontal Design (Horizontal-MRM)*

In this design method, which is very similar to that presented by [1], the full load operation of the prime mover extends until the yearly PES becomes greater than a minimum PES, instead of being positive (Figure 6.5). The prime mover size remains the same as the prime mover size recommended by the classic MRM.

Deriving the equation for the above condition results in the constraint presented in Eq. (6-10):

$$PES = \frac{\sum\limits_{yearly} F_{pp} + \sum\limits_{yearly} F_b - \sum\limits_{yearly} F_{CCHP}}{\sum\limits_{yearly} F_{pp} + \sum\limits_{yearly} F_b} \geq PES_{min} \tag{6-8}$$

$$\left(1 - PES_{min}\right) \sum\limits_{yearly} \frac{\eta_{e,CCHP} F_{CCHP}}{\eta_{e,pp} \eta_g} - \sum\limits_{yearly} F_{CCHP} + \left(1 - PES_{min}\right) \sum\limits_{yearly} \frac{\zeta}{\eta_b} \geq 0 \tag{6-9}$$

and finally

$$\sum\limits_{yearly} F_{CCHP} \leq \frac{1 - PES_{min}}{\eta_b \left[1 - \left(1 - PES_{min}\right) \frac{\eta_{e,CCHP}}{\eta_{e,pp} \eta_g}\right]} \sum\limits_{yearly} \zeta \tag{6-10}$$

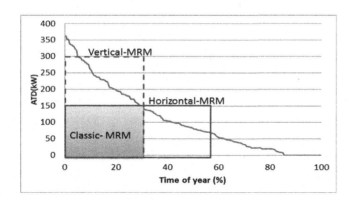

Figure 6.5 Graphical representation of horizontal-, and vertical-MRM with the constraint of PES > PES$_{min}$.

The FLO and PLO efficiency of components is assumed to be equal in Eqs. (6-8) to (6-10).

The yearly fuel consumption of the CCHP system ($\sum_{yearly} F_{CCHP}$) is dependent on the operation mode of the prime mover. When the prime mover operates in full load it consumes fuel according to the nominal efficiency, but when it operates in partial load it consumes more fuel and follows its partial load efficiency. Hence

$$\sum_{yearly} F_{CCHP} = \sum_{FLO} F_{CCHP} + \sum_{PLO} F_{CCHP} = \sum_{FLO} \frac{E_{nom}}{\eta_{e,CCHP}^{FLO}} + \sum_{PLO} \frac{E_{PM}}{\eta_{e,CCHP}^{PLO}} \tag{6-11}$$

$$E_{PM} = \begin{cases} E_{dem} & E_{dem} \leq E_{nom} \\ E_{nom} & E_{dem} > E_{nom} \end{cases} \tag{6-12}$$

Depending on the type of prime mover, its electrical efficiency in FLO and PLO conditions ($\eta_{e,CCHP}^{FLO}$, $\eta_{e,CCHP}^{PLO}$) can be found from the data presented in Chapters 1 and 2.

6.3.2 Vertical Design (Vertical-MRM)

In vertical-MRM, the full load operation time of the prime mover is the same as that recommended by the classic MRM, but the prime mover size becomes greater until the PES remains greater than a minimum PES. Figure 6.5 depicts a graphical explanation of this method.

The equation that can be derived for the vertical-MRM is similar to that derived for the horizontal-MRM in Eqs. (6-10) to (6-12).

In the MRM methods the length of full load operation time may be known (the classic MRM) or unknown (the horizontal-MRM) but the problem is determining the time when the prime mover is supposed to operate in full load or partial load. To determine the time when the prime mover should be programmed for FLO or PLO another constraint should be introduced. This constraint can be based on the cooling, heating, or electrical loads of the consumer. The constraint can also be expressed based on different economic criteria. Moreover, the ability to sell electricity to the grid has an impact on the time for FLO and PLO. In addition, different fuel and electricity tariffs in peak demand and nonpeak demand times can change the constraint. Specifying the constraint depends on the decision-making of the designer. The decision should be made based on evaluations and calculations such as those mentioned in this paragraph.

6.3.3 High-level Analysis

While energy has different aspects such as thermodynamic (energy and exergy), economic, and environmental, both the vertical- and horizontal-MRM sizing

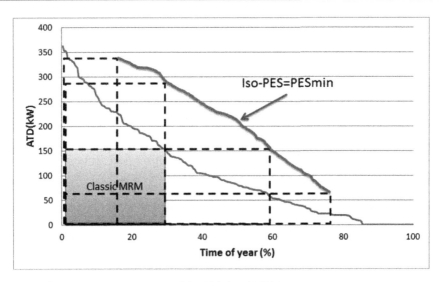

Figure 6.6 Graphical representation of the high-level MRM.

methods use only the energy criterion to size the prime mover. The high-level MRM uses the economic criterion of net present value (NPV) after applying the PES constraint. In this method, by changing the prime mover size and full load operation time, the pairs (prime mover size and full load operation time) that result in the same PES_{min} are found and an iso-PES line is drawn. Then in the second step the NPV criterion is applied to the sizes related to the iso-PES line to find the pair that results in the maximum NPV. The pair with the maximum NPV would be the answer. A graphical representation of the high-level MRM is depicted in Figure 6.6.

The advantage of the high-level MRM is that among different options with the same primary energy savings, the one that creates more economic benefit is chosen as the solution.

6.4 EMS Sizing Methods

Energy management strategies (EMSs) have been used to find better solutions in the sizing of prime movers for CCHP systems [3-9]. The most popular EMS sizing methods include FTL, FEL, and following the seasonal load (FSL). Commonly, researchers have used different strategies to size a CCHP system and then have compared the results from different points of view. The final result depends on the comparisons, investor priorities, and the designer decisions. In the following the details of these sizing methods are introduced.

6.4.1 FTL Sizing

In this method the prime mover size is chosen to provide the maximum thermal demand from the heat recovery of the CCHP system. This means that in case of lack/surplus of electricity it would be purchased/sold from/to the grid respectively. Since the thermal demand changes, the recovered heat from the prime mover may exceed the thermal demand in some conditions. In such conditions the extra heat may be stored in a heat storage system or lost. When working with EMSs, usually FTL is more popular because the extra electricity or lack of electricity can be sold to or bought from the grid and the prime mover also can be programmed to provide the heat demand (no heat lost or storage). In fact in this method the prime mover's main priority is to produce heat, and electricity is a byproduct of the CCHP system.

EMSs only pay attention to the thermal or electrical loads of the consumer; they pay no attention to other aspects of energy such as the economic and environmental criteria. These criteria may be considered only when comparing different EMS sizing methods to select the best energy management strategy.

A formula for the FLT sizing method is presented here:

$$E_{nom}^{FTL} = \left\{ E_{nom} \mid (B = 0 \vee Q_{rec} = ATD_{max}) \right\} \tag{6-13}$$

where E_{nom}^{FTL} is the nominal size of the prime mover recommended by the FTL sizing method, and B is the auxiliary boiler size to compensate for the lack of heat for different purposes.

6.4.2 FEL Sizing

In the FEL sizing method the prime mover size is chosen to provide the maximum electrical demand from the prime mover of the CCHP system. This means that when there is a lack of heat it will be compensated for by an auxiliary heating system such as a water boiler. Since in the FEL sizing method the electricity demand is the main priority of the CCHP system in case of surplus recoverable heat, it is recommended to store the heat in a heat storage system for reuse when needed; otherwise it would be wasted.

When the prime mover is programmed to operate in partial load, no surplus electricity is produced, but if it is supposed to operate in full load, since the electricity demand is changeable, the electricity demand may be less than the full load electricity production of the prime mover. In this case the electricity may be sold to the grid, or be consumed by the auxiliary electrical cooling or heating systems integrated into the CCHP system. Electricity storage systems (batteries) can also be used to store the surplus electricity and sell it to the grid during peak hours when electricity is more expensive. In addition by applying some optimization algorithms the prime mover size can be smaller than the maximum electrical demand and instead electricity storage systems can be used to satisfy the peak demand. This helps to reduce fuel consumption and pollution production, but it may cost more.

The prime mover size in FEL sizing (E_{nom}^{FEL}) is calculated as follows:

$$E_{nom}^{FEL} = E_{dem,\max} \tag{6-14}$$

6.4.3 FSL Sizing

In this method, the prime mover size is designed to follow thermal and electrical loads depending on the monthly or seasonal electricity to heat demand ratio (LR) of the consumer:

$$LR = \frac{Monthly\ E_{dem}}{Monthly\ H_{dem}} \tag{6-15}$$

If $LR > 1$, the prime mover should be programmed to run in FTL strategy in that month, otherwise it should be programmed to work in FEL strategy. According to this discussion, the prime mover size can be calculated as follows:

$$E_{nom}^{FSL} = \max(E_{nom}^{FEL}, E_{nom}^{FTL}) \tag{6-16}$$

6.5 Thermodynamic Sizing Methods

Thermodynamic sizing methods use energy and exergy analyses. In these methods usually the primary energy consumption (PEC) and overall efficiency of the CCHP system are considered as the first law of thermodynamics criteria; also the second law efficiency (π) and exergy destruction ($\dot{I}$) of the CCHP system are considered as the second law of thermodynamics criteria. In order to find the optimum size of the CCHP systems from the energy point of view the Eqs. (6-17) and (6-18) must be maximized:

$$FESR = \frac{\displaystyle\sum_{yearly} PEC_{SCHP} - \sum_{yearly} PEC_{CCHP}}{\displaystyle\sum_{yearly} PEC_{SCHP}} \tag{6-17}$$

$$\eta_o = \frac{\displaystyle\sum_{yearly}(E_{PM} + E_g + H_{dem} + C_{dem} + D_{dem})}{\displaystyle\sum_{yearly} PEC_{CCHP}} \tag{6-18}$$

where

$$E_g = \begin{cases} 0 & if\ \ E_{PM} \ge E_{dem} \\ E_{dem} - E_{PM} & if\ \ E_{PM} < E_{dem} \end{cases} \tag{6-19}$$

where PEC is the fuel energy consumption to provide all energy demand types for the consumer. E_{PM} and E_g are the electricity produced by the prime mover and electricity purchased from the grid, respectively. Using FESR as the energy criterion is common in the research; we also recommend using this criterion since it compares the fuel consumption of the CCHP with the SCHP system as well.

In order to find the optimum CCHP size from the exergy point of view the Eqs. (6-20) and (6-21) must be maximized:

$$EXIR = \frac{\pi_{CCHP} - \pi_{SCHP}}{\pi_{CCHP}} \tag{6-20}$$

$$\dot{I}RR = \frac{\dot{I}_{SCHP} - \dot{I}_{CCHP}}{\dot{I}_{SCHP}} \tag{6-21}$$

where EXIR and $\dot{I}RR$ stand for exergy efficiency increase ratio, and exergy destruction reduction ratio, respectively. In addition π and $\dot{I}$ should be calculated yearly as follows:

$$\pi = \frac{\sum\limits_{yearly} \dot{\phi}_{out}}{\sum\limits_{yearly} \dot{\phi}_{in}} \tag{6-22}$$

$$\dot{I} = \sum\limits_{yearly} \dot{\phi}_{in} - \sum\limits_{yearly} \dot{\phi}_{out} \tag{6-23}$$

where $\dot{\phi}_{in}$ and $\dot{\phi}_{out}$ are the supplied and recovered exergy rate. Guides for calculation of ϕ are given in Chapter 3.

Exergy and energy criteria usually result in different optimum sizes for the CCHP system. Therefore, if we are supposed to use both energy and exergy criteria simultaneously, we must compromise between the two points of view in order to propose a single optimum size.

A solution for this problem is to introduce another function that is a linear combination of the energy and exergy criteria. For example, if we have decided to use FESR and EXIR simultaneously for sizing, the following equation can be maximized:

$$\begin{aligned} ENEX &= \omega_{EN} FESR + \omega_{EX} EXIR \\ \omega_{EN} &+ \omega_{EX} = 1 \\ 0 &\leq \omega_{EN}, \ \omega_{EX} \leq 1 \end{aligned} \tag{6-24}$$

where ω_{EN} and ω_{EX} are the energy and exergy weights, respectively, and $ENEX$ stands for energy/exergy function.

Another solution is to determine the optimum size recommended by every criterion independently and then combine the optimum sizes linearly as follows:

$$E_{nom}^{ENEX} = \omega_{EN} E_{nom}^{FESR} + \omega_{EX} E_{nom}^{EXIR}$$
$$\omega_{EN} + \omega_{EX} = 1 \tag{6-25}$$
$$0 \le \omega_{EN}, \ \omega_{EX} \le 1$$

The CCHP size proposed by Eqs. (6-24) and (6-25) usually maximizes neither FESR nor EXIR; in fact it is a compromise between energy and exergy criteria.

While calculating the thermodynamic criteria, attention must be paid to the operation mode of the prime mover. According to the analyses one decide to operate it in partial load, full load, or a combination of PLO and FLO during the course of a year.

In addition it should be fully understood that it is unfeasible to reuse all of the recoverable heat from the prime mover due to the highly changeable heating and cooling demands. Therefore any calculation should be done according to the energy demands of the consumer. It is evident that the selling or saving of the surplus electricity produced by the prime mover (if it exists) should be decided according to the economic evaluations. Thermodynamic sizing methods cannot help in decisions about such problems alone. This weakness of thermodynamic analyses is eliminated in the thermoeconomical sizing methods presented in the following.

6.6   Thermoeconomic Sizing Methods

In thermoeconomic analyses, in addition to thermodynamic analyses, the economic aspect of energy and energy conversion equipment is also considered. In the literature review presented in Chapter 1 many economic criteria that have been used by researchers were introduced. Table 1.1 summarized the economic criteria into four categories of present value methods, rate of return methods, ratio methods, and payback methods. The most commonly used criteria from each category are the net present value (NPV), internal rate of return (IRR), premium value percentage (PVP), and payback period (PB).

The NPV calculates the real value of the project after its lifetime (L), in other words it tells us if the project will be profitable during its lifetime or not. NPV can be calculated as follows:

$$NPV = -I + \frac{SV}{(1+r)^L} + \sum_{y=1}^{L} \frac{cf_y}{(1+r)^y} \tag{6-26}$$

where r is the interest rate, and I is the initial capital cost of the CCHP system and includes every cost before operation such as the costs of a consultant, hardware, transportation, installation, tests, site rent, labor, tax, etc.

cf_y is the annual net cash, the annual summation of every positive and negative cash flow after operation during the lifetime of the project ($cf_y = (er - ex)_y$). The positive cash flow, which means earnings (er) includes the yearly costs that are not paid in comparison with the SCHP system, such as the electricity provided by the CCHP system, the heating, DHW, or cooling provided by heat recovery, and electricity sold back to the grid. The negative cash flow mean expenses (ex), including yearly costs paid for operation, maintenance, and meeting consumer demands. These expenses may include the costs of fixed and variable O&M, taxes, fuel, and electricity purchased from the grid. SV is the salvage value of the project after its lifetime.

Positive NPV means profitability and in the sizing process the CCHP size with the maximum NPV is the most profitable CCHP.

According to Eq. (6-26) the bigger the annual net cash flow (cf_y) is, the more positive NPV would be. In addition a smaller initial capital cost (I) increases the NPV.

Due to the importance of cf_y some designers [3, 10–13] have preferred to optimize this parameter or its components such as annual costs, avoided costs, annual savings, etc.

In the NPV methods attention is not paid to the interest rate and economic situation of the country where the investment takes place. The same projects respond differently from the economic point of view in countries with different economic situations. For this reason, there must be a tool to evaluate the risk of investment. IRR is very useful for this purpose. IRR is the interest rate at which the profitability of the project is zero. The bigger the IRR is, the safer the investment will be. IRR can be calculated according to the following equation:

$$-I + \frac{SV}{(1+IRR)^L} + \sum_{y=1}^{L} \frac{cf_y}{(1+IRR)^y} = 0 \qquad (6\text{-}27)$$

A CCHP size is profitable when IRR is greater than the interest rate r. The larger the profitability margin (IRR $- r$) is, the lower the risk of investment will be. In unstable economics, a minimum profitability margin should be considered to decrease the risk of investment.

If we are required to know the net profit for every dollar of investment in different sizes of CCHP system, the premium value percentage (PVP) should be calculated as follows:

$$PVP = \frac{NPV}{I} \qquad (6\text{-}28)$$

The CCHP size with the highest PVP is the best. For calculation of the PVP, the NPV must be positive.

Another criterion that is easily understandable, especially for nonexpert investors, is the payback period (PB). This criterion calculates the number of years it takes to recover the original capital cost. It is important to mention that this criterion does not calculate the years it takes to recover the *real value* of the investment cost, because it does not consider the interest rate. It means that if you have invested 1 USD today you will receive exactly 1 USD after the payback period. This criterion is sometime misleading because it may calculate the payback period while the NPV is negative. This is due to neglecting the impact of interest rate in calculating the PB. PB can be calculated as follows:

$$PB = \frac{I}{\overline{cf}} \qquad (6\text{-}29)$$

where $\overline{cf} = \frac{1}{L}\sum_{y=1}^{L} cf_y$ is the yearly average net cash flow. In order to avoid a misleading PB and incorrect decision-making, we propose a modified payback period equation:

$$PB' = L\frac{I}{NPV + I} = L\frac{1}{PVP + 1} \qquad (6\text{-}30)$$

This equation predicts the payback period more realistically than Eq. (6-29) because it considers the real value of the profit (NPV) in the calculations.

In addition to the above criteria, levelized electricity production cost (LEPC) can also be considered as a design criterion. This criterion calculates the cost per kilowatt of electricity production in the CCHP system. A smaller LEPC means a more efficient CCHP system thermoeconomically. LEPC can be calculated according to Eq. (1-17).

6.7   Multicriteria Sizing Methods

Since energy has different aspects, optimizing one aspect may decrease the benefits of others. In addition, when we need to consider other characteristics of energy and CCHP systems such as the environmental impact, the problem will become more complex and designing a CCHP system with optimum design criteria from different points of view becomes unfeasible. Due to the importance of different characteristics of energy and CCHP systems, none of the criteria can be omitted or neglected; therefore a solution should be proposed to consider all of these characteristics. Researchers have proposed the fitness function (ff) and multicriteria sizing function (MCSF). In the following sections these methods are further explained and expanded upon in the context of sizing CCHP systems.

6.7.1   Fitness Function Sizing Method

References [11] and [14] used a fitness function to design CCHP systems. A sample of the ff is presented in Eq. (1-36). By looking at this equation we can see that FESR, annual total cost savings (ATCS), and CO_2 reduction ratio are considered as the nominated criteria of the thermodynamic, economic, and environmental aspects of the energy and CCHP system, respectively. In addition, an equal weight is given to the three criteria. To expand upon this method, we propose considering some criteria that each have several subcriteria. In addition, the weight of every criterion and subcriterion can be calculated using weighting methods such as AHP or the fuzzy method presented in Chapter 4. The fitness function is formulated as follows:

$$ff = \sum_{i=1}^{n} w_i ff_i,$$
$$\sum_{I=1}^{n} w_i = 1 \tag{6-31}$$

where

$$ff_i = \sum_{j=1}^{k_i} \beta_{ij} w_{ij} C_{ij}$$
$$\sum_{i=1}^{n} \sum_{j=1}^{k_i} w_{ij} = 1 \tag{6-32}$$
$$\beta_{ij} = \begin{cases} 1 & \textit{if the } C_{ij} \textit{ is considered in the analyses} \\ 0 & \textit{if the } C_{ij} \textit{ is not considered in the analyses} \end{cases}$$

where n is the number of criteria and k_i is the number of subcriteria of the ith criterion. In addition w_i is the weight of ith criterion and w_{ij} is the weight of the jth subcriterion of the ith criteria. C_{ij} is the normalization of the jth subcriterion of the ith criteria. According to the normalization method, the optimization may be done by maximizing or minimizing the ff. The C_{ij} is defined as follows:

$$C_{ij} = \frac{c_{ij}^{CCHP} - c_{ij}^{SCHP}}{c_{ij}^{CCHP}}, \quad c_{ij} > 0 \tag{6-33}$$

where c_{ij} is a profit such as the exergy efficiency where higher the better.

If c_{ij} is a cost such as the fuel consumption where smaller the better, it would be normalized as follows:

$$C_{ij} = \frac{c_{ij}^{SCHP} - c_{ij}^{CCHP}}{c_{ij}^{SCHP}}, \quad c_{ij} > 0 \tag{6-34}$$

Equation (6-31) can be optimized by using different optimization algorithms such as the genetic algorithm. As mentioned previously, the weights of the criteria and subcriteria should be calculated using the weighting methods presented in Chapter 4.

It is important to mention that optimizing the fitness function does not guarantee the optimization of every criteria or subcriteria. In fact the weakness of this method is that the optimum ff may result in some nonoptimum criteria or subcriteria, even those that are most important for the designer. This is because when the ff is being optimized we have no control on the variation of ff_i and C_{ij}.

Another weakness of this method is that some of the subcriteria may be steadily increasing or decreasing. Such subcriteria may cause the ff to become steadily increasing or decreasing. In such cases, since we optimize all of the subcriteria together, all of the subcriteria should be checked individually and we should find the subcriterion (or criteria) that caused this divergence. The designer may make different decisions about the subcriteria that are to blame; for example their weight can be set to zero, another subcriterion can be used instead, another sizing method can be used, etc.

6.7.2 Multicriteria Sizing Function

The MCSF proposed by [15] optimizes every subcriterion of C_{ij} individually and finds the optimum CCHP size proposed by the subcriterion. Then after giving weight to every criterion and subcriterion, a linear combination of weights and optimum CCHP sizes proposed by the sub-criteria is calculated, and the prime mover size of the CCHP system is recommended as follows:

$$E_{nom}^{MCSF} = \sum_{i=1}^{n} \sum_{j=1}^{k_i} \beta_{ij} \omega_i \omega_{ij} E_{opt}^{ij} \qquad (6\text{-}35)$$

$$E_{opt}^{ij} = \left\{ E_{nom} \mid C_{ij}(E_{nom}, Conditions) \big|_{opt}, i=1,\ldots,n \ \& \ j=1,2,\ldots,k_i \right\} \qquad (6\text{-}36)$$

where E_{opt}^{ij} is the optimum CCHP size, or the proposed CCHP size under some conditions and constraints from the jth sub-criterion of the ith criterion, and E_{nom}^{MCSF} is the nominal electrical capacity of the prime mover proposed by the MCSF.

As we discussed previously, in problems such as CCHP sizing where many criteria are encountered, it is basically impossible to find a CCHP size to optimize all of the criteria and subcriteria simultaneously. Therefore we should look for a compromise or quasi-optimum solution. In comparison with the ff method, the MCSF has absolute control over the optimization of every subcriterion. Therefore, E_{nom}^{MCSF} tends to approach the size of E_{opt}^{ij} for which the corresponding criterion and subcriterion is the most important (has the biggest $\omega_i \omega_{ij}$) for the designer. In addition, since all of the subcriteria are optimized individually, if there is a subcriterion that is steadily increasing or decreasing, we can use our engineering feelings about the problem and according to the trend of the subcriterion reach an agreement about the recommended engine size from that subcriterion. This means we may use the optimum sizes of some criteria and the agreed sizes proposed by some nonoptimizable subcriteria in the MCSF. In other words, with the MCSF there is always a solution.

6.8   Case Study

Design a CCHP system to provide the energy demands of the building described in
the previous chapter. Use MRM, EMS, *ff*, and MCSF, and compare the results of
the methods. In addition, show the impact of climate difference on the design of the
CCHP cycle for the five climates presented in Chapter 5. Assume that the prime mover is operating at full load the whole year and that the DHW system is prepared by heat
recovery. The economic, environmental, and technical input data for the written code
are presented in Appendix 2 [13, 16-33].

6.8.1   Solution Outline

In order to design a CCHP system, first we should decide upon the prime mover type.
The prime mover was previously chosen based on the multicriteria decision-making
method in Chapter 4 for the five climates. The fuzzy logic and grey incidence approaches
recommended using an internal combustion (IC) engine for all five climates. In the second step, the energy demands of the residential building were calculated and presented in
Chapter 5. In the third step, the components (apart from the internal combustion engine)
that are supposed to be used in the CCHP cycle should be introduced as a basic CCHP
system. Since the exhaust gas of an IC engine is about 540 °C, this energy source can be
used for cooling production in single-effect absorption chillers. The water is preheated
by the lube oil cooler and engine water jacketing before being heated by the exhaust
gases. If the water temperature is high enough for the chiller to cover the cooling load (or
heating load), the temperature control valve (TCV) bypasses the hot water to the down
stream of auxiliary boiler and it enters the chiller (or heating system), otherwise it will be
reheated by the auxiliary heater. The CCHP cycle consisting of the IC engine, absorption
chiller, auxiliary heater, and heating system are presented in Figure 6.7. Determining the

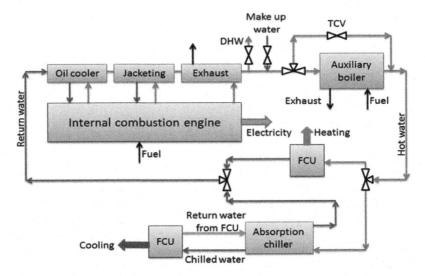

Figure 6.7 CCHP cycle components.

prime mover size dictates the size of the auxiliary boiler. In addition, the chiller size is determined according to the maximum cooling load. Therefore the IC engine should be sized first.

6.8.2 Sizing Using the MRM

The ATD of the building is first determined using the MRM method. The ATD curves of the building for the five climates are depicted in Figure 6.8. Since the load calculation was done based on the TNM, the horizontal axis of ATD curves is presented in percent time of year.

After determining the ATD, the rectangle with maximum area and FLO time should be found for every climate. For our purposes, the area of the rectangle is calculated against the time of year in Figure 6.9. The IC engine size can be found by using the maximum area and its corresponding FLO time in Eq. (6-3). The results of sizing using the MRM are presented in Figure 6.10.

6.8.3 Sizing Using EMS Methods

EMS sizing methods including the FTL, FEL, and FSS can also be used for sizing the IC engine according to Eqs. (6-13), (6-14), and (6-16). In these methods it is enough to determine the peak of ATD and the electrical demand for every climate. In the FTL

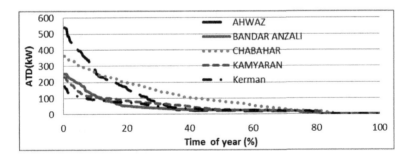

Figure 6.8 The ATD of the building in the five climates.

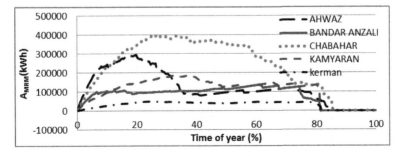

Figure 6.9 The area of the rectangle based on the MRM method versus time of year.

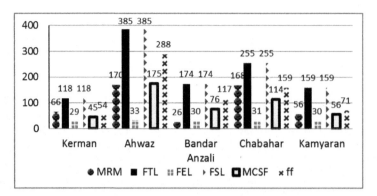

Figure 6.10 The IC engine size proposed by different sizing methods.

sizing method the engine size should be chosen to provide as much recoverable heat as ATD_{max}. Therefore the correlations that were presented for Q_{rec} and E_{nom} of the IC engine in the second chapter can be used to calculate the engine size. For example, Eq. (2-10) is used in this example since the electrical demand is below 100 kW. As an example the ATD_{max} of Ahwaz is 541.40 kW, therefore by using Eq. (2-10) we have $541.40 = 1.368E_{nom} + 14.57$ or $E_{nom} = 385.11$ kW. The results of the EMS for the five climates are presented in Figure 6.10 and compared with the MRM results.

As can be seen among the sizing methods, the FTL and FSL approaches recommend the biggest engine size. A bigger engine size means a higher initial investment cost and more heat loss since the recoverable heat high enough to cover the maximum aggregated thermal demand (it means heat is wasted when the ATD decreases). In addition, operation and maintenance of a bigger engine costs more. The positive side of a bigger engine is production of surplus electricity to be sold to the grid and no need for an auxiliary boiler. These can improve the economic benefits of the CCHP system. As can be seen many parameters are interacting with each other and the final decision can only be made when all of the vital criteria and subcriteria are calculated. As opposed to the FTL and FSL, the engine sizes proposed by the FEL are small. A small engine size has its own characteristics as well. For example, smaller initial capital outlay, and lower operation and maintenance costs are the advantages, while the need for an auxiliary boiler and less excess electricity to be sold to the grid can be mentioned as disadvantages. In contrast to the EMS sizing methods, in most cases MRM proposes intermediate engine sizes with respect to those proposed by the FTL, FSL, and FEL sizing methods. The intermediate engine size benefits from the advantages of small and big engine sizes, however again the final decision can only be made when all of the essential criteria are calculated. This means that using the MRM or EMS sizing methods without including thermodynamic, economic, and environmental criteria is risky.

Another weakness of the MRM and EMS sizing methods is that they are completely building-demand oriented and do not consider the impact from the components of the CCHP system. For example if using MRM or EMS as sizing methods there will be no difference between the size of the prime mover of a basic CCHP and a CCHP system that is integrated with solar heating or a thermal storage system.

6.8.4 Sizing Using ff

In order to use *ff* for sizing the IC engine, the criteria, subcriteria, and their weights should be known.

In this example we consider FESR and EXIR as the thermodynamic subcriteria; NPV, IRR and PB as the economic subcriteria; and the reduction ratio of CO_2, CO, and NO_x as the environmental subcriteria. The AHP weighting method is used and most of the pairwise comparisons presented in Chapter 4 (Tables 4.7A, B, and D) can be used again except for the economic subcriteria (Table 4.7C) due to consideration of different subcriteria. The pairwise comparison of economic subcriteria and weights are presented in Tables 6.1 and 6.2, respectively.

After calculating the weights, the subcriteria should be calculated according to Eqs. (6-33) and (6-34). Among the subcriteria, FESR, CO2RR, CORR, and NOXRR should be normalized according to Eq. (6-34) and EXIR should be normalized according to Eq. (6-33). The economic subcriteria, however, cannot be normalized using Eqs. (6-33) and (6-34). This is because the net annual cash flow (cf_y) is always negative for the SCHP system, and as a result the NPV, IRR, and PB would become negative (meaningless). This is because we receive no economic benefit from the SCHP system. For this reason NPV, IRR, and PB are normalized as follows:

$$C_{31} = \begin{cases} c_{31}^{CCHP}/\left(c_{31}^{CCHP} - c_{31}^{SCHP}\right), & if\ c_{31}^{CCHP} > 0 \\ 0, & if\ c_{31}^{CCHP} \leq 0 \end{cases} \tag{6-37}$$

Table 6.1 **Pairwise Comparison of Economic Subcriteria**

	NPV	IRR	PB
NPV	JE	WMI	SMI(R)
IRR	WMI(R)	JE	VSMI(R)
PB	SMI	VSMI	JE

Table 6.2 **Weights of Criteria and Subcriteria Considered for Sizing**

Criteria	Weight	Subcriteria	Weight
Thermodynamic	0.34	FESR (c_{11})	0.55
		EXIR(c_{12})	0.45
Environmental	0.20	CORR(c_{21})	0.29
		CO2RR(c_{22})	0.33
		NOXRR(c_{23})	0.38
Economic	0.46	NPV(c_{31})	0.33
		IRR(c_{32})	0.50
		PB(c_{33})	0.17

$$C_{32} = \begin{cases} \left(c_{32}^{CCHP} - r\right)/c_{31}^{CCHP}, & \text{if } c_{32}^{CCHP} > r \\ 0, & \text{if } c_{32}^{CCHP} \le r \end{cases} \tag{6-38}$$

$$C_{33} = \begin{cases} 1/c_{33}^{CCHP}, & \text{if } 0 < \overline{cf} < I_{CCHP} \\ 1, & \text{if } I_{CCHP} < \overline{cf} \\ 0, & \text{if } \overline{cf} \le 0 \end{cases} \tag{6-39}$$

To calculate the C_{ij}, we should calculate the c_{ij} first. For this purpose in the following all of the subcriteria are formulated according to the cycle presented.

The first subcriterion is the fuel consumption of the CCHP and SCHP systems. To present the following equations in a compact form, whenever j is used as a superscript it can take two values of $j = 0$ or $j = 1$, in which 0 represents the SCHP system and 1 stands for a CCHP system. In addition an energy vector including four components (E, C, H, and D) is used for the electricity, cooling, heating, and DHW demands. This vector is also used for other subcriteria. Hence

$$F^j = (F_E^j, F_C^j, F_H^j, F_D^j), \ j = 0, 1 \tag{6-40}$$

$$c_{11}^j = \sum_{yearly} (F_E^j + F_C^j + F_H^j + F_D^j) \tag{6-41}$$

where the fuel vector components are calculated as follows:

$$\begin{aligned}
F_E^0 &= \frac{E_{dem}}{\eta_{pp}\eta_g}, & F_E^1 &= \frac{E_{PM} + Q_{rec}}{\eta_{CHP}} + \frac{(E'_{dem})}{\eta_{pp}\eta_g} \\
F_C^0 &= \frac{C_{dem}}{\eta_b COP_{abc}\varepsilon_{FCU}}, & F_C^1 &= \frac{C'_{dem}}{\eta_b COP_{abc}\varepsilon_{FCU}} \\
F_H^0 &= \frac{H_{dem}}{\eta_b\varepsilon_{FCU}}, & F_H^1 &= \frac{H'_{dem}}{\eta_b\varepsilon_{FCU}} \\
F_D^0 &= \frac{D_{dem}}{\eta_{wh}}, & F_D^1 &= \frac{D_{dem}}{\eta_{wh}}
\end{aligned} \tag{6-42}$$

where

$$E_{PM} = \begin{cases} E_{nom}, & \text{if } FLO \\ E_{dem}, & \text{if } E_{nom} \geq E_{dem} \wedge PLO \\ E_{nom}, & \text{if } E_{nom} < E_{dem} \wedge PLO \end{cases}$$

$$Q_{rec} = \begin{cases} 1.368 E_{PM} + 14.57, & 30 \leq E(kW) \leq 500 \\ 1.854 E_{PM}, & 0 \leq E(kW) < 30 \end{cases}$$

$$E'_{dem} = \begin{cases} E_{dem} - E_{PM}, & \text{if } E_{dem} > E_{PM} \\ 0, & \text{if } E_{dem} \leq E_{PM} \end{cases} \tag{6-43}$$

$$C'_{dem} = \begin{cases} C_{dem} - COP_{abc} Q_{rs}, & \text{if } C_{dem} > COP_{abc} Q_{rs} \\ 0, & \text{if } C_{dem} \leq COP_{abc} Q_{rs} \end{cases}$$

$$H'_{dem} = \begin{cases} H_{dem} - Q_{rs}, & \text{if } H_{dem} > Q_{rs} \\ 0, & \text{if } H_{dem} \leq Q_{rs} \end{cases}$$

$$Q_{rs} = Q_{rec} + Q_{solar}$$

Since the DHW energy demand is calculated separately, the CCHP system can be designed with a separate DHW system (sprt-DHW) or be integrated with the heating system and use the recovered heat (intg-DHW). In this example the intg-DHW system is used and the DHW energy demand is combined with the heating energy demand. The prime (′) in the above equations represents the additional energy required (heat from the auxiliary boiler or electricity from the grid) to provide the corresponding energy demand for the building. Q_{solar}, which is the solar energy collected by a solar water heater, will be calculated in the next chapter.

Since the engine is supposed to operate at full load all time, the recoverable heat may exceed the heat demand at some particular times. The extra heat will be wasted because no heat storage is considered to store the surplus heat. In order to calculate the annual energy loss of the CCHP system the following equation can be used:

$$ATD \leq Q_{rs} \rightarrow H_{loss}^{CCHP} = Q_{rs} - ATD \tag{6-44}$$

Additionally, sometimes the heat demand (for cooling or heating) may exceed the recoverable heat. In this case the lack of heat should be compensated for by burning fuel in the auxiliary boiler and water heater (for the sprt-DHW). The annual lack of heat that should be provided by systems other than the solar and heat recovery systems can be calculated as follows:

$$ATD \geq Q_{rs} \rightarrow ATD' = H'_{dem} + C'_{dem} / COP_{abc} + D_{dem} \tag{6-45}$$

After calculation of fuel consumption for the CCHP and SCHP systems, the FESR can be calculated according to Eq. (6-34).

In the exergy analysis, the supplied exergy to ($\dot{\phi}_{in}$) and recovered exergy from ($\dot{\phi}_{out}$) the CCHP and SCHP systems are used to calculate the exergy efficiency (π) as follows:

$$\dot{\phi}_{in}^{j} = (\dot{\phi}_{in,E}^{j}, \dot{\phi}_{in,C}^{j}, \dot{\phi}_{in,H}^{j}, \dot{\phi}_{in,D}^{j})$$

$$\dot{\phi}_{out}^{j} = (\dot{\phi}_{out,E}^{j}, \dot{\phi}_{out,C}^{j}, \dot{\phi}_{out,H}^{j}, \dot{\phi}_{out,D}^{j})$$

(6-46)

$$\pi^{j} = \sum_{yearly} (\dot{\phi}_{out}^{j}) \bigg/ \sum_{yearly} (\dot{\phi}_{in}^{j})$$

(6-47)

The exergy vector components are calculated according to the cycle as follows:

$$\dot{\phi}_{in,E}^{0} = F_{E}^{0} \chi(T_{hg}) - (F_{E}^{0} - E_{dem}) \chi(T_{amb})$$

$$\dot{\phi}_{in,C}^{0} = \dot{W}_{sp} + \dot{W}_{rp} + \dot{W}_{wp} + F_{C}^{0} \chi(T_{hg}) - F_{C}^{0} (1-\eta_{b}) \chi(T_{amb})$$

$$\dot{\phi}_{in,H}^{0} = \dot{W}_{wp} + F_{H}^{0} \chi(T_{hg}) - F_{H}^{0} (1-\eta_{b}) \chi(T_{amb})$$

$$\dot{\phi}_{in,D}^{0} = F_{D}^{0} \chi(T_{hg}) - F_{D}^{0} (1-\eta_{b}) \chi(T_{amb})$$

(6-48)

$$\dot{\phi}_{out,E}^{0} = E_{dem}$$

$$\dot{\phi}_{out,C}^{0} = C_{dem} \chi(T_{room})$$

$$\dot{\phi}_{out,H}^{0} = H_{dem} \chi(T_{room})$$

$$\dot{\phi}_{out,D}^{0} = D_{dem} \chi(T_{room})$$

(6-49)

$$\dot{\phi}_{in,E}^{1} = F_{E}^{1} \chi(T_{hg}) - (F_{E}^{1} - E_{PM} - Q_{rec}) \chi(T_{amb}) + E_{dem}'$$

$$\dot{\phi}_{in,C}^{1} = \dot{W}_{sp} + \dot{W}_{rp} + \dot{W}_{wp} + F_{C}^{1} \chi(T_{hg}) - F_{C}^{1} (1-\eta_{b}) \chi(T_{amb})$$

$$\dot{\phi}_{in,H}^{1} = \dot{W}_{wp} + F_{H}^{1} \chi(T_{hg}) - F_{H}^{1} (1-\eta_{b}) \chi(T_{amb})$$

$$\dot{\phi}_{in,D}^{1} = F_{D}^{1} \chi(T_{hg}) - F_{D}^{1} (1-\eta_{b}) \chi(T_{amb})$$

(6-50)

$$\dot{\phi}_{out,E}^{1} = E_{PM} + E_{dem}' + \dot{\phi}_{Q_{rs}}$$

$$\dot{\phi}_{out,C'}^{1} = C_{dem}' \chi(T_{room})$$

$$\dot{\phi}_{out,H'}^{1} = H_{dem}' \chi(T_{room})$$

$$\dot{\phi}_{out,D}^{1} = D_{dem} \chi(T_{room})$$

(6-51)

where ambient temperature is calculated according to the weather information presented in Chapter 5:

$$
T_{amb} = \begin{cases} T_{db,sum} & if \ (H_{dem} = 0 \wedge C_{dem} \neq 0) \\ T_{db,win} & if \ (C_{dem} = 0 \wedge H_{dem} \neq 0) \\ T_{room} & if \ H_{dem} = C_{dem} = 0 \end{cases} \tag{6-52}
$$

$$
T_{room} = \begin{cases} T_{room,sum} & if \ (C_{dem} = 0 \wedge H_{dem} \neq 0) \\ T_{room,win} & if \ (H_{dem} = 0 \wedge C_{dem} \neq 0) \\ T_{amb} & if \ H_{dem} = C_{dem} = 0 \end{cases} \tag{6-53}
$$

$$
\chi(T_x) = 1 - T_0 / T_x
$$

Since Q_{rs} comprises the recovered heat from the oil cooler, water jacketing, exhaust, and solar system, its exergy is calculated as follows:

$$
\begin{aligned} Q_{rs} &= Q_{oil} + Q_{jacketing} + Q_{exhaust} + Q_{solar} \rightarrow \dot{\phi}_{Q_{rs}} = Q_{oil} \chi(T_{oil}) \\ &+ Q_{jacketing} \chi(T_{jacketing}) + Q_{exhaust} \chi(T_{exhaust}) + Q_{solar} \chi(T_p) \end{aligned} \tag{6-54}
$$

After calculating the exergy efficiency the EXIR can be calculated according to Eq. (6-33).

In order to investigate the environmental impact of the CCHP with respect to the SCHP system the pollution product by the energy demand components should be evaluated. The pollution vector is defined as follows:

$$
X^j = (X_E^j, X_C^j, X_H^j, X_D^j) \tag{6-55}
$$

where X can be CO, CO_2, or NO_x. The emission production of each pollutant also is determined as follows:

$$
Em_X^j = \sum_{yearly} (X_E^j + X_C^j + X_H^j + X_D^j)_t \tag{6-56}
$$

The components of the pollution vector are calculated as follows:

$$
\begin{aligned} X_E^0 &= E_{PM} \cdot i_{X,E}^0, \ X_C^0 = C_{dem} \cdot i_{X,C}^0 \\ X_H^0 &= H_{dem} \cdot i_{X,H}^0, \ X_D^0 = D_{dem} \cdot i_{X,D}^0 \end{aligned} \tag{6-57}
$$

$$X_E^1 = E_{PM} \cdot i_{X,E}^1 + E'_{dem} \cdot i_{X,E}^0, X_C^1 = C'_{dem} \cdot i_{X,C}^1$$
$$X_H^1 = H'_{dem} \cdot i_{X,H}^1, X_D^1 = D_{dem} \cdot i_{X,D}^1 \tag{6-58}$$

where i is the pollution index based on kg/MWh. After calculation of Em_X^j for each pollutant, the emission reduction ratio can be calculated according to Eq. (6-34).

In the economic evaluations, the initial investment cost and net annual cash flow are the key parameters. Initial investment cost is the summation of equipment costs. The capital cost vector and investment cost are presented as follows:

$$I^j = (I_E^j, I_C^j, I_H^j, I_D^j) \tag{6-59}$$

$$I^j = I_E^j + I_C^j + I_H^j + I_D^j \tag{6-60}$$

The capital vector components are calculated as follows:

$$I_E^0 = NOU.i_E^0, I_E^1 = E_{nom} \cdot i_E^1 + I_E^0 \tag{6-61}$$

$$I_C^0 = C_{nom} \cdot i_C^0, I_C^1 = C_{nom} \cdot i_C^1$$
$$C_{nom} = \max(C_{dem}) \tag{6-62}$$

$$I_H^0 = B^0 \cdot i_H^0, B^0 = \max \max(H_{dem}, C_{dem} / COP_{abc})$$
$$I_H^1 = B^1 \cdot i_H^1 + A_{co} \cdot i_{solar}^1, B^1 = \max \max(H'_{dem}, C'_{dem} / COP_{abc}) \tag{6-63}$$

$$I_D^0 = I_D^1 = \begin{cases} NOU \times i_D & sprt - DHW \\ 0 & \mathrm{int} g - DHW \end{cases} \tag{6-64}$$

where i is the price index in currency-unit/kW (for example USD/kW) except for i_E^0 which is in currency-unit/NOU. NOU means the number of units in the building (each unit has an electricity counter) to measure the electricity flow from the grid to the building. B, C_{nom}, and A_{co} are the nominal capacity of the boiler, chiller, and the solar collector area, respectively.

The net annual cash flow calculation starts with determining the earnings (positive cash flow) and expenses (negative cash flow). The earning (er) and expense (ex) vectors and the net annual cash flow (cf_y) are presented here:

$$er^j = (er_E^j, er_C^j, er_H^j, er_D^j), ex^j = (ex_E^j, ex_C^j, ex_H^j, ex_D^j)$$
$$er^j = er_E^j + er_C^j + er_H^j + er_D^j, ex^j = ex_E^j + ex_C^j + ex_H^j + ex_D^j \tag{6-65}$$

$$cf_y^j = -I_{OM}^j + (er^j - ex^j)_y \qquad (6\text{-}66)$$

where I_{OM}^j is the yearly operation and maintenance costs. The components of the earning and expense vectors are formulated below:

$$
\begin{aligned}
er_E^0 &= 0,\; ex_E^0 = E_{dem} \cdot t_{be} \\
er_C^0 &= 0,\; ex_C^0 = F_C^0 \cdot t_{gas} \\
er_H^0 &= 0,\; ex_H^0 = F_H^0 \cdot t_{gas} \\
er_D^0 &= 0,\; ex_D^0 = F_D^0 \cdot t_{gas}
\end{aligned}
\qquad (6\text{-}67)
$$

$$
er_E^1 =
\begin{cases}
E_{dem} \cdot t_{be} + (E_{PM} - E_{dem}) \cdot t_{se} & E_{PM} \geq E_{dem} \\
E_{PM} \cdot t_{be} & E_{PM} < E_{dem}
\end{cases}
$$

$$
ex_E^1 = \left(F_E^1 - \frac{(E_{dem}')}{\eta_{pp}\eta_g} \right) \cdot t_{gas} + E_{dem}' \times t_{be}
$$

$$
\begin{aligned}
er_C^1 &= (F_C^0 - F_C^1) \cdot t_{gas},\; ex_C^1 = F_C^1 \cdot t_{gas} \\
er_H^1 &= (F_H^0 - F_H^1) \cdot t_{gas},\; ex_H^1 = F_H^1 \cdot t_{gas} \\
er_D^1 &= 0,\; ex_D^1 = F_D^1 \cdot t_{gas}
\end{aligned}
\qquad (6\text{-}68)
$$

where t is a tariff and the subscripts of gas, se, and be stand for natural gas, selling electricity, and buying electricity. E_{PM} is the electrical output of the engine and should not be mistaken with E_{nom}.

After calculation of I and cf_y the economic criteria of NPV, IRR, and PB can be calculated according to Eqs. (6-26), (6-27), and (6-29), respectively. Also these criteria can be normalized according to Eqs. (6-37) to (6-39), respectively.

Up to now we have formulated all of the subcriteria. By combining these criteria and the corresponding weights in the fitness function (Eq. 6-31) the magnitude of ff can be drawn against the engine nominal size (E_{nom}) as in Figure 6.11. As can be seen in Figure 6.11 the ff for all five climates has an optimum point and reaches a maximum. The optimum sizes for the five climates are given in Figure 6.10 for comparison with other sizing methods. In addition, the results presented in figure 6.12 shows that every criterion for the climate of Kamyaran has an optimum point. According to the sizes proposed by ff and comparing them with the sizes proposed by the MRM and EMS for the climate of Kamyaran, it can be seen from Figure 6.13 that using MRM and FEL give smaller subcriteria in most cases except for C_{22} and C_{11}. In addition, in comparison with the ff, using FTL and FSL results in smaller or equal values for all of the subcriteria. It should be noted that using the MRM or EMS sizing methods may result in economically unbeneficial CCHP systems (see the C_{31}, C_{32} for Kamyaran when using the FEL sizing method). Furthermore, Figure 6.13 reveals that EXIR has no optimum point and is increasing steadily, but fortunately combining it with FESR results in ff_1, which has an optimum point (Figure 6.12).

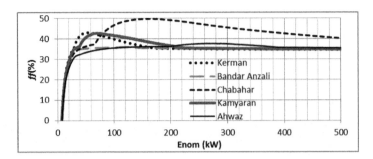

Figure 6.11 The variation of *ff* versus engine nominal size for the five climates.

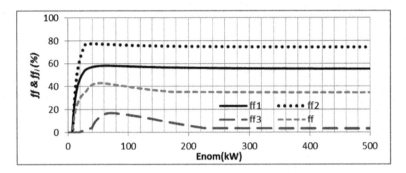

Figure 6.12 *ff* and *ff_i* for the climate of Kamyaran.

6.8.5   Sizing Using MCSF

The MCSF and *ff* methods have many parts in common except in the method they use to calculate the optimum size. The subcriteria calculated in the *ff* method are used to calculate the optimum or proposed engine size for every subcriterion. An example of the subcriterion is depicted for the climate of Kamyaran in Figure 6.13. As can be seen, except for C_{12} there are optimum sizes for each subcriterion. Regarding C_{12}, according

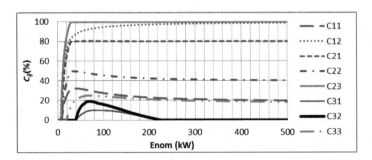

Figure 6.13 Variation of the C_{ij} versus engine size for the climate of Kamyaran.

to the trend of EXIR we have assumed an EXIR of 90% as the design constraint for this subcriterion. After calculating the engine sizes proposed by the subcriteria, they are combined with the weights in Eq. (6-35) to calculate the final size (Figure 6.10). In comparison with the *ff*, the sizes proposed by the MCSF are smaller in all climates. Smaller engine sizes mean a smaller initial investment. This smaller capital cost does not guarantee the profitability of the CCHP system. This should be checked according to the subcriteria curves that are presented in Figures 6.14 to 6.21. In addition the

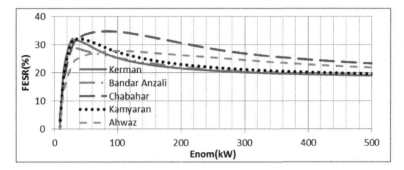

Figure 6.14 C_{11} variation with respect to E_{nom} in different climates.

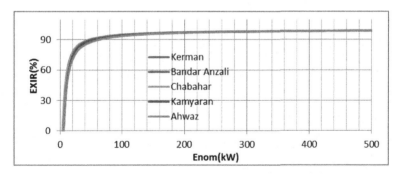

Figure 6.15 C_{12} variation with respect to E_{nom} in different climates.

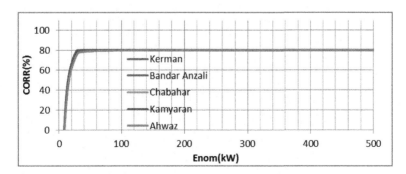

Figure 6.16 C_{21} variation with respect to E_{nom} in different climates.

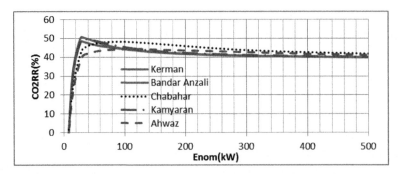

Figure 6.17 C_{22} variation with respect to E_{nom} in different climates.

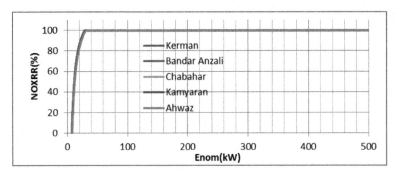

Figure 6.18 C_{23} variation with respect to E_{nom} in different climates.

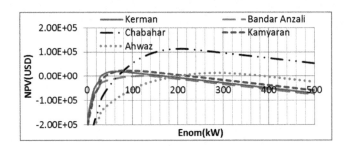

Figure 6.19 NPV variation with respect to E_{nom} in different climates.

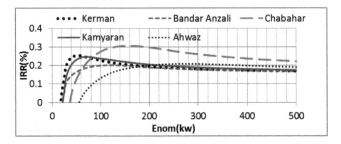

Figure 6.20 IRR variation with respect to E_{nom} in different climates.

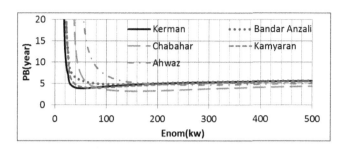

Figure 6.21 PB variation with respect to E_{nom} in different climates.

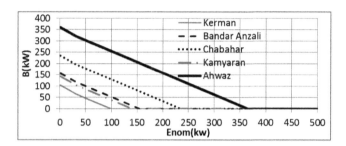

Figure 6.22 Boiler size with respect to the engine size.

Table 6.3 The Absorption Chiller Boiler Size for Different Climates Using MCSF

	Kerman	Ahwaz	Bandar Anzali	Chabahar	Kamyaran
Chiller size (kW)	105	361	159	236	145
Boiler size (kW)	52	183	76	117	81

profitability of the sizes that were proposed by other sizing methods can be checked with these curves. The chiller size only depends on the maximum cooling load for every climate, but the boiler size depends on the engine size and building demands. The boiler size is plotted against E_{nom} in Figure 6.22, which can be used for both *ff* and MCSF. Table 6.3 presents the boiler and chiller sizes when designing with MCSF.

6.9 Problems

1. Use the building demands calculated in previous chapter problems to calculate the ATD of the building.
2. By using the MRM design a CCHP system for your case study.
3. Use horizontal-MRM and vertical-MRM to design the CCHP system and compare the results from the economic point of view.

4. What would be the size of the engine if you use high-level analysis instead of MRM?
5. Design the CCHP system using the EMS.
6. Use MCSF and *ff* to design the CCHP for the building you have considered.
7. If the CCHP is supposed to use a combination of electrical and absorption chillers, modify the formulations for this change.
8. If the CCHP is supposed to use a thermal storage system, draw the schematic of the new CCHP system.
9. Modify the formulations for the new CCHP system you have drawn in the Problem 8.
10. The price of electricity is usually higher at peak hours. Evaluate if it is profitable to use an electricity storage system to store electricity at low load hours and sell it in peak hours or not. Calculate the PB and NPV of this idea.

References

[1] Cardona, E., Piacentino, A., 2003. A Methodology for Sizing a Trigeneration Plant in Mediterranean Areas. Applied Thermal Engineering 23, 1665–1680.
[2] Martinez-Lera, S., Ballester, J., 2010. A Novel Method for the Design of CHCP (Combined Heat, Cooling and Power) Systems for Buildings. Energy 35, 2972–2984.
[3] Mago, P.J., Chamra, L.M., 2009. Analysis and Optimization of CCHP Systems Based on Energy, Economical, and Environmental Considerations. Energy and Buildings 41, 1099–1106.
[4] Cho, H., Mago, P.J., Luck, R., Chamra, L.M., 2009. Evaluation of CCHP Systems Performance Based on Operational Cost, Primary Energy Consumption, and Carbon Dioxide Emission by Utilizing an Optimal Operation Scheme. Applied Energy 86, 2540–2549.
[5] Mago, P.J., Hueffed, A.K., 2010. Evaluation of a Turbine Driven CCHP System for Large Office Buildings under Different Operating Strategies. Energy and Buildings 42, 1628–1636.
[6] Wang, J.-J., Jing, Y.-Y., Zhang, C.-F., (John) Zhai, Z., 2011. Performance Comparison of Combined Cooling Heating and Power System in Different Operation Modes. Applied Energy 88, 4621–4631.
[7] Liu, M., Shi, Y., Fang, F., 2012. A New Operation Strategy for CCHP Systems with Hybrid Chillers. Applied Energy 95, 164–173.
[8] Gu, Q., Ren, H., Gao, W., Ren, J., 2012. Integrated Assessment of Combined Cooling Heating and Power Systems under Different Design and Management Options for Residential Buildings in Shanghai. Energy and Buildings 51, 143–152.
[9] Chen, X.P., Wang, Y.D., Yu, H.D., Wu, D.W., Li, Yapeng, Roskilly, A.P., 2012. A Domestic CHP System with Hybrid Electrical Energy Storage. Energy and Buildings 55, 361–368.
[10] Li, C.Z., Shi, Y.M., Huang, X.H., 2008. Sensitivity Analysis of Energy Demands on Performance of CCHP System. Energy Conversion and Management 49, 3491–3497.
[11] Wang, J.-J., Jing, Y.-Y., Zhang, C.-F., 2009. Optimization of Capacity and Operation for CCHP System by Genetic Algorithm. Applied Energy 87, 1325–1335.
[12] Lozano, M.A., Carvalho, M., Serra, L.M., 2009. Operational Strategy and Marginal Costs in Simple Trigeneration Systems. Energy Xxx, 1–8.
[13] Sanaye, S., Ardali, M.R., 2009. Estimating the Power and Number of Microturbines in Small-Scale Combined Heat and Power Systems. Applied Energy 86, 895–903.
[14] Wang, J., (John) Zhai, Z., Jing, Y., Zhang, C., 2010. Particle Swarm Optimization for Redundant Building Cooling Heating and Power System. Applied Energy 87, 3668–3679.

[15] Ebrahimi, M., Keshavarz, A., 2013. Sizing the Prime Mover of a Residential Micro-CCHP System by Multi-Criteria Sizing Method for Different Climates. Energy 54, 291–301.

[16] Wu, D.W., Wang, R.Z., 2006. Combined Cooling, Heating and Power: A Review. Progress in Energy and Combustion Science 32, 459–495.

[17] Kong, X.Q., Wang, R.Z., Huang, X.H., 2004. Energy Efficiency and Economic Feasibility of CCHP Driven by Stirling Engine. Energy Conversion and Management 45, 1433–1442.

[18] Katsigiannis, P.A., Papadopoulos, D.P., 2005. A General Technoeconomic and Environmental Procedure for Assessment of Small-Scale Cogeneration Scheme Installations: Application to a Local Industry Operating in Thrace, Greece, Using Microturbines. Energy Conversion and Management 46, 3150–3174.

[19] Onovwiona, H.I., Ugursal, V.I., 2006. Residential Cogeneration Systems: Review of the Current Technology. Renewable and Sustainable Energy Reviews 10, 389–431.

[20] Yagoub, W., Doherty, P., Riffat, S.B., 2006. Solar Energy-Gas Driven Micro-CHP System for an Office Building. Applied Thermal Engineering 26, 1604–1610.

[21] Godefroy, J., Boukhanouf, R., Riffat, S., 2007. Design, Testing and Mathematical Modelling of a Small-Scale CHP and Cooling System (Small CHP-Ejector Trigeneration). Applied Thermal Engineering 27, 68–77.

[22] Huangfu, Y., Wu, J.Y., Wang, R.Z., Kong, X.Q., Wei, B.H., 2007. Evaluation and Analysis of Novel Micro-Scale Combined Cooling, Heating and Power (MCCHP) System. Energy Conversion and Management 48, 1703–1709.

[23] Pehnt, M., 2008. Environmental Impacts of Distributed Energy Systems—The Case of Micro Cogeneration. Environmental Science & Policy 11, 25–37.

[24] Wang, J.-J., Jing, Y.-Y., Zhang, C.-F., Shi, G.-H., Zhang, X.-T., 2008. A Fuzzy Multi-Criteria Decision-Making Model for Trigeneration System. Energy Policy 36, 3823–3832.

[25] Wang, J.-J., Jing, Y.-Y., Zhang, C.-F., Zhang, X.-T., Shi, G.-H., 2008. Integrated Evaluation of Distributed Triple-Generation Systems Using Improved Grey Incidence Approach. Energy 33, 1427–1437.

[26] Kuhn, V., Klemes, J., Bulatov, I., 2008. MicroCHP: Overview of Selected Technologies, Products and Field Test Results. Applied Thermal Engineering 28, 2039–2048.

[27] Sugiartha, N., Tassou, S.A., Chaer, I., Marriott, D., 2009. Trigeneration in Food Retail: An Energetic, Economic and Environmental Evaluation for a Supermarket Application. Applied Thermal Engineering 29, 2624–2632.

[28] Ren, H., Zhou, W., Nakagami, K., Gao, W., 2010. Integrated Design and Evaluation of Biomass Energy System Taking into Consideration Demand Side Characteristics. Energy 35, 2210–2222.

[29] Tichi, S.G., Ardehali, M.M., Nazari, M.E., 2010. Examination of Energy Price Policies in Iran for Optimal Configuration of CHP and CCHP Systems Based on Particle Swarm Optimization Algorithm. Energy Policy 38, 6240–6250.

[30] Monteiro, E., Afonso Moreira, N., Ferreira, S., 2009. Planning of Micro-Combined Heat and Power Systems in the Portuguese Scenario. Applied Energy 86, 290–298.

[31] U.S. Environmental Protection Agency, http://www.epa.gov/ttnchie1/ap42/ch01/related/c01s04.html.

[32] Vice-Presindency for Strategic Planning and Supervision, http://www.spac.ir.

[33] U.S. Energy Information Administration, www.eia.gov.

CCHP Solar Heat Collectors

7

7.1 Introduction

Integrating renewable energy resources such as solar energy with the high efficiency energy conversion equipment such as CCHP systems amplify the benefits of these technologies. Solar energy systems can be used as photovoltaic cells to provide solar electricity, as heating collectors for heating [1-8], or as photovoltaic/thermal (PVT) systems for combined production of heat and electricity [9]. Solar collectors appear in different shapes and types such as plate, concentric, parabolic, evacuated tube, etc. [10, 11]. Solar thermal energy and recovered heat from the CCHP system can be combined to provide as much as possible of the aggregated thermal demand. They also can be used as the running energy of ORC-CCHP cycles. Some of the CCHP systems that have used solar technologies were presented in Figures 1.5, 1.14, 1.21, 1.22, and 2.32.

Solar system selection depends on the heat demand, temperature required, and application. The most common type for commercial, residential, and low-temperature industrial applications is the plate collector. This type of collector can provide air or hot water at a temperature of up to 93 °C but its efficiency decreases rapidly below 50% at temperatures higher than 70 °C. These collectors usually are installed in a fixed position. Wolf GmbH is a main producer of this type of collector [12]. Concentrating collectors are also available for high-temperature industrial applications and provide hot water with temperatures higher than 115 °C. These collectors need a sun tracer to receive as much solar heat as possible. [11]. Chromasun, Inc. has produced a micro concentrator (MCT) that is able to produce hot water with a temperature of up to 200 °C while maintaining an efficiency higher than 50% [13]. Figure 7.1 compares the MCT working temperature range and efficiency with different types of collectors. The collector efficiency is defined as the ratio of useful heat gained by the collector per unit of aperture area (W/m^2) to the total irradiation of the collector (W/m^2).

In order to design an optimum hybrid fossil fuel/solar-driven CCHP system two important steps must be taken.

First, the direction and orientation of the solar collector must be optimized to receive the maximum solar heat at that position. Theoretically, we can find an optimum orientation for the solar collector at every time and location to receive the maximum solar heat at that time and location. However, in practice usually a year is divided into some limited number of periods and an optimum orientation for each period can be calculated. For example, we can design one optimum direction to receive the maximum yearly solar energy, or four optimum orientations for the four seasons. This means that the collector orientation must be adjusted for every season according to its corresponding optimum orientation. When heat gain and optimum direction of the collector are determined, the size of the collector should be optimized in the second step. Collector size should be chosen so that the extra cost due to the solar system is

Combined Cooling, Heating and Power.

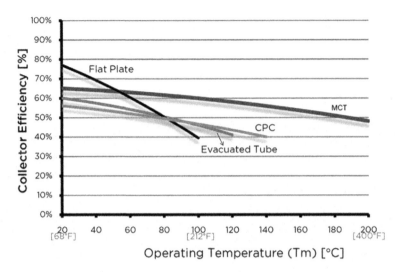

Figure 7.1 Comparison of the efficiency of different collectors at $T_{amb} = 20$ °C [13].

recovered by decreasing the engine and auxiliary boiler size as well as by decreasing fuel consumption and environmental pollution.

Since the MRM and EMS sizing methods are independent from component type and size, these methods cannot be used for designing a hybrid fossil fuel/solar CCHP system. The methods that best suit this problem are the fitness function (ff) and multicriteria sizing function (MCSF). These methods are influenced by consumer demands and component type and size as well. In the following the calculation of the solar heat gain by a plate collector is presented and then it is shown how the optimum direction can be found. In addition, the size of the solar collector is calculated and we discuss how it can be determined according to the engine size and building demand.

7.2 Solar Heat Gain Calculation [10, 11, 14, and 15]

Among different types of solar collectors the most common type, the plate collector, is considered for coupling with a CCHP system. In addition, two options, single glazing and double glazing, are considered. The total heat gain of a plate collector with area A_{co} is defined as follows:

$$Q_{solar} = q_u.A_{co} \tag{7-1}$$

In the above equation q_u is the collector heat index in watts per square meter of the collector area. In this section was want to find the magnitude of the collector heat index in the optimum direction.

The collector heat index is the maximum average heat that the collector absorbs during a period of time in the optimum orientation. The axis on which the Earth spins is tilted 23.45°. In order to find the collector heat index, the declination angle δ must

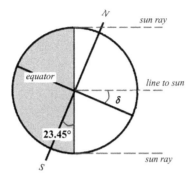

Figure 7.2 Representation of the declination angle.

be calculated for each day of year. This angle is shown in Figure 7.2. As can be seen it is the angle between the sun rays and equator, and is calculated as follows:

$$\delta = 23.45 \sin\left(360° \times \frac{284 + N}{365} \right) \tag{7-2}$$

where N is the day number starting with $N = 1$ for January 1st and ending with $N = 365$ for December 31st, and annual changes of δ are negligible. The sun orientation is determined by solar azimuth angle az_s in the horizontal plane (angle HOS), and the solar altitude σ (angle HOQ in Figure 7.3).

The solar angular hour AH is also calculated as follows:

$$\begin{aligned} AH &= (\textit{Number of hours from solar noon}) \times 15° \\ &= \quad \left|12 - \textit{time in hour}\right| \times 15° \end{aligned} \tag{7-3}$$

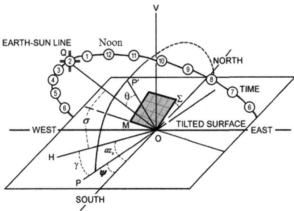

Figure 7.3 Sun orientation with respect to the collector at point O on earth (θ, az_s, σ, ψ and γ are the incident, solar azimuth, solar altitude, surface azimuth, and surface-solar azimuth angles, respectively).

Therefore, σ and az_s are calculated as follows:

$$\sin \sigma = \cos(LAT).\cos(\delta).\cos(AH) + \sin(LAT).\sin(\delta) \tag{7-4}$$

$$\sin az_s = \cos(\delta).\sin(AH)/\cos \sigma \tag{7-5}$$

$$\cos az_s = \frac{\sin \sigma.\sin(LAT) - \sin(\delta)}{\cos \sigma.\cos(LAT)} \tag{7-6}$$

The angle between the sun array attacking earth at point O (OQ), and the perpendicular axis to the collector (OP') is called the incident angle θ and is calculated as follows:

$$\cos \theta = \cos \sigma.\cos \gamma.\sin \Sigma - \sin \sigma.\cos \Sigma \tag{7-7}$$

where Σ is the tilting angle of the collector and γ is the collector-sun azimuth angle. When the collector is facing southeast γ is

$$\gamma = \begin{cases} az_s - \psi & \text{morning} \\ az_s + \psi & \text{afternoon} \end{cases} \tag{7-8}$$

where ψ is the collector azimuth angle; if the collector faces southwest γ is

$$\gamma = \begin{cases} az_s + \psi & \text{morning} \\ az_s - \psi & \text{afternoon} \end{cases} \tag{7-9}$$

For calculation of the total solar heat gained during each day of a year, the day length (DL) sunrise time, and sunset time are required and are calculated as follows:

$$DL = \frac{2}{15} \arccos(-\tan(LAT).\tan(\delta)) \tag{7-10}$$

$$sunrise = 12 - 0.5 DL$$
$$sunset = sunrise + DL \tag{7-11}$$

The direct solar radiation on the Earth's surface with a cloudless and clean sky is calculated by the following equation:

$$RI_{DN} = Ae^{-B/\sin \sigma} \tag{7-12}$$

where A and B are apparent extraterrestrial irradiation and atmospheric extinction coefficient, respectively. They are functions of the date and take into account the seasonal variation of the Earth/Sun distance and the air's water vapor content. The magnitudes of A and B are curve fitted according to the data presented in Ref. [10] as follows:

$$A = a_1 .e^{-\left(\frac{N-b_1}{c_1}\right)^2} + a_2 .e^{-\left(\frac{N-b_2}{c_2}\right)^2}$$
$$a_1 = 1187; b_1 = 371.9; c_1 = 225.9; a_2 = 1144$$
$$b_2 = -6.401; c_2 = 206.2$$

(7-13)

$$B = a_6 .e^{-\left(\frac{N-b_6}{c_6}\right)^2} + a_7 .e^{-\left(\frac{N-b_7}{c_7}\right)^2} + a_3 .e^{-\left(\frac{N-b_3}{c_3}\right)^2}$$
$$a_6 = 0.007325; b_6 = 235; c_6 = 22.82; a_7 = 0.07095$$
$$b_7 = 184.90; c_7 = 99.03; a_3 = 0.1409; b_3 = 783.3; c_3 = 3624$$

(7-14)

The total solar irradiation $I_{t\theta}$ of a collector for any direction and tilting angle Σ with an incidence angle of θ is measured as follows:

$$RI_{t\theta} = RI_{DN} .\cos\theta + RI_{d\theta} + RI_{re}$$

(7-15)

where $RI_{DN} \cos\theta$ is the direct irradiation component, $RI_{d\theta}$ is the diffusion coming from the sky, and RI_{re} is the reflected short wave irradiation from the foreground that possibly reaches the collector. To estimate the diffuse component, dimensionless parameter C is defined; it depends on the dust and moisture content of the atmosphere and changes through the year. It is defined by the following equation:

$$C = \frac{RI_{dH}}{RI_{DN}}$$

(7-16)

where RI_{dH} is the diffuse component on a horizontal surface on a cloudless day. The following equation is used to estimate the diffuse irradiation on a collector with a tilting angle of Σ:

$$RI_{d\theta} = C.RI_{DN} .F_{ss}, \quad F_{ss} = \frac{1 + \cos\Sigma}{2}$$

(7-17)

where C is curve fitted using the data presented in Ref. [10], as follows:

$$C = a_4 .e^{-\left(\frac{N-b_4}{c_4}\right)^2} + a_5 .e^{-\left(\frac{N-b_5}{c_5}\right)^2}$$
$$a_4 = 0.08041; b_4 = 187.4; c_4 = 85.71; a_5 = 0.05888$$
$$b_5 = 179.3; c_5 = 589.6$$

(7-18)

The reflected component can be calculated as follows:

$$RI_{re} = RI_{tH} \cdot \rho_g \cdot F_{sg}, \; F_{sg} = \frac{1 - \cos \Sigma}{2} \tag{7-19}$$

where ρ_g is the reflectance coefficient and RI_{tH} is the total radiation on a horizontal surface. Bituminous surfaces reflect less than 10% of the total solar irradiation [10]. For simplicity in this investigation it is assumed $RI_{re} = (RI_{DN} \cos \theta + RI_{d\theta})/9$. If the place where the collector is installed is something other than bituminous, the reflectance should be modified. Finally the heat gained by the collector is calculated as follows:

$$q_u = RI_{t\theta}(\tau\alpha)_\theta - U_L(T_p - T_{amb}(N, HR)) \tag{7-20}$$

To calculate the solar heat at every hour, the ambient temperature should be given for every hour (HR) of each day (N). The data from Ref. [10] are curve fitted to calculate τ, α, and U_L as follows:

$$\tau = \begin{cases} p_1\theta^4 + p_2\theta^3 + p_3\theta^2 + p_4\theta + p_5, \text{single glazing} \\ p_6\theta^6 + p_7\theta^5 + p_8\theta^4 + p_9\theta^3 + p_{10}\theta^2 + p_{11}\theta + p_{12}, \text{double glazing} \end{cases}$$

$$p_1 = -5.084 \times 10^{-8}; p_2 = 5.098 \times 10^{-6}; p_3 = -17.57 \times 10^{-5}$$

$$p_4 = 19.28 \times 10^{-4}; p_5 = 86.83 \times 10^{-2}; p_6 = 5.764e \times 10^{-11} \tag{7-21}$$

$$p_7 = -1.388e \times 10^{-8}; p_8 = 1.2e \times 10^{-6}; p_9 = -4.696 \times 10^{-5}$$

$$p_{10} = 79.92 \times 10^{-5}; p_{11} = -46.22 \times 10^{-4}; p_{12} = 77.04 \times 10^{-2}$$

$$\alpha = q_1\theta^8 + q_2\theta^7 + q_3\theta^6 + q_4\theta^5 + q_5\theta^4 + q_6\theta^3 + q_7\theta^2 + q_8\theta + q_9$$

$$q_1 = -6.2 \times 10^{-15}; q_2 = 4.873 \times 10^{-13}; q_3 = 1.416 \times 10^{-10}$$

$$q_4 = -2.479 \times 10^{-8}; q_5 = 1.575 \times 10^{-6}; q_6 = -4.776 \times 10^{-5} \tag{7-22}$$

$$q_7 = 65.62 \times 10^{-5}; q_8 = -31.26 \times 10^{-4}; q_9 = 0.96$$

$$U_{32} = 0.03083T_p + 5.517, \; U_{-12} = 0.025T_p + 5.5, \text{single glazing} \tag{7-23}$$

$$U_{32} = 0.01733T_p + 3, \; U_{-12} = 0.01556T_p + 2.678, \text{double glazing} \tag{7-24}$$

$$U_L = U_{-12} + \frac{(U_{32} - U_{-12})(T_{amb}(N, HR) + 12)}{44} \tag{7-25}$$

In order to find the optimum collector direction to receive the maximum average annual solar heat the following procedure should be followed.

1. After calculation of q_u at every hour of daytime during a year, the yearly average value of q_u for every pair of (Σ, ψ) should be calculated $(\overline{q}_{u, yearly})$.
2. The angle pair (Σ, ψ) in which $\overline{q}_{u, yearly}$ is maximum should be determined as $(\Sigma_{opt}, \psi_{opt})$.
3. The maximum of $\overline{q}_{u, yearly}$ is the collector heat index and is shown by $q_{s,opt} = \overline{q}_{u, yearly, max}$, hence

$$q_{s,opt} = \max\left(\frac{\displaystyle\sum_{N=1}^{365}\sum_{sunrise}^{sunset} q_u\big|_{(\Sigma, \psi)}}{\displaystyle\sum_{N=1}^{365} DL}\right) \quad 0 \leq \Sigma, \psi \leq 90° \tag{7-26}$$

7.3 Collector Size

After calculation of the collector heat index, the size of the collector should be determined. Since the solar energy is used in parallel with the heat recovery from the prime mover, the summation of these two heat sources is responsible for providing the heating, cooling, DHW, or ATD. For this purpose four strategies can be considered as follows:

ATD strategy: The collector is designed to provide ATD_{max}, therefore

$$A_{co} = (ATD_{max} - Q_{rec, min})/q_{s,opt} , \quad ATD_{max} \geq Q_{rec, min} \tag{7-27}$$

H strategy: The collector is designed to provide the maximum heating load demand $(H_{dem, max})$, therefore

$$A_{co} = (H_{dem, max} - Q_{rec, min})/q_{s,opt} , \quad H_{dem, max} \geq Q_{rec, min} \tag{7-28}$$

C strategy: The collector is designed to fulfill the maximum cooling load $(C_{dem, max})$, therefore

$$A_{co} = \frac{C_{dem, max} - COP_{abc} \cdot Q_{rec, min}}{COP_{abc} \cdot q_{s,opt}}, \quad C_{dem, max} \geq COP_{abc} \cdot Q_{rec, min} \tag{7-29}$$

D strategy: The collector is designed to provide $D_{dem, max}$, therefore

$$A_{co} = D_{dem, max}/q_{s,opt} \tag{7-30}$$

When DHW is integrated into the CCHP system (intg-DHW) the first three strategies are applicable; if DHW is separated from the CCHP system (sprt-DHW) the last three strategies are applicable. Choosing the best strategy for collector sizing depends of the thermodynamic, environmental, and economic evaluations. In this study DHW is integrated into the CCHP system.

With regard to collector sizing strategies it should be noted that the maximum cooling, heating, and ATD remains constant, but the recoverable heat from the engine is changeable due to possible changing of the prime mover size in the optimization process. Therefore the collector size changes with the size of the engine and the optimum collector size for the CCHP will be chosen according to the evaluation criteria of the CCHP cycle.

7.4 Case Study

In this chapter we will use solar thermal energy in the CCHP system that was designed in the previous chapter. The cycle is depicted in Figure 7.4.

In the natural gas/solar-driven CCHP an internal combustion engine is used to produce electricity, as chosen in Chapter 4. As can be seen a branch of water that is supposed to be heated for production of cooling, heating, or DHW is heated through the solar system and the rest of the water flows through the heat recovery exchangers to capture the heat from the lube oil, water jacketing, and exhaust gases. Then they mix together and enter the auxiliary boiler; if needed the boiler starts heating water to reach an appropriate energy level according to the building demands, otherwise a temperature control valve (TCV) bypasses the water to down stream of the boiler. The output of the auxiliary boiler can enter a fan coil unit (FCU) for heating purposes or the absorption chiller to produce chilled water. The chilled water is sent to the FCU for cooling purposes.

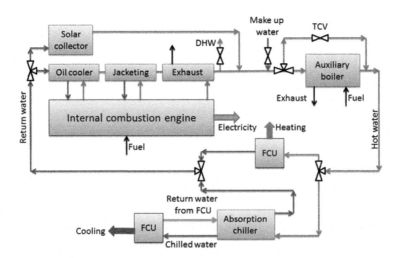

Figure 7.4 Schematic of the natural gas/solar-driven CCHP.

DHW is extracted from the pipe just after the heat recovery exchangers, and the make-up water is also supplied to the main water line just before the auxiliary boiler to avoid overheating in the boiler. Since the solar system provides heat during the daytime we expect decreased fuel consumption from the auxiliary boiler and decreased fuel cost accordingly. In addition decreasing the fuel consumption decreases environmental pollutions as well. It should be noted that different designers may consider different configurations for the solar collector and the basic CCHP system depending on the temperature that the solar collector can produce.

In the first step the optimum direction of the collector is determined for the five climates. In addition the results are compared for single- and double-glazed collectors.

Figure 7.5 presents the annual average solar heat gain in the climate of Kamyaran when the collector faces southwest. The collector is double-glazed. As can be seen an optimum value happens at a particular angle pair of (ψ, Σ). The graphs of other climates and directions follow the same pattern and to avoid repetition only Figure 7.5 is presented, but the final results are tabulated in Table 7.1. As this table shows, the optimum direction for the five climates is southwest with different angles of (ψ, Σ). In addition the double-glazed collector receives more heat than the single-glazed type. Among the five climates, the climate of Ahwaz benefits most from the solar energy and receives 327.03 W/m^2. Bandar Anzali receives the least solar energy at 255.95 W/m^2.

In the second step, the collector size must be determined. As discussed in Eqs. (7-27) to (7-30) the collector size depends on the engine size and maximum of H_{dem}, C_{dem}, DHW, and ATD. This dependency is demonstrated in Figure 7.6 for the climate of Kamyaran. The other climates follow a similar pattern, therefore to avoid repetition only the results for Kamyaran are presented. As can be seen the collector area decreases as the engine size increases and finally in a particular engine size the collector area becomes zero. This means that since it is assumed that the engine is operating at full load, the recoverable heat fulfills the cooling, heating, and DHW demands of the

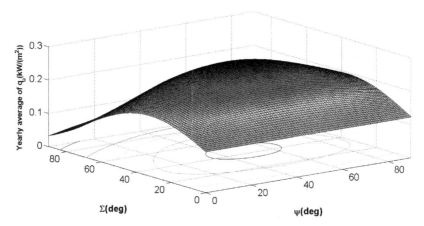

Figure 7.5 q_u for different angles of ψ and Σ in the southwest direction for double-glazed collector, Kamyaran.

Table 7.1 Optimum Direction, Type, and Collector Heat Index in Five Climates

	Single-Glazed, Southeast			Double-Glazed, Southeast		
	$q_{s,opt}$	ψ_{opt}	Σ_{opt}	$q_{s,opt}$	ψ_{opt}	Σ_{opt}
Kerman	82.72	0	26	196.30	0	29
Ahwaz	128.05	0	27	227.91	0	29
Bandar Anzali	62.87	0	33	176.22	0	35
Chabahar	114.33	0	23	225.20	0	25
Kamyaran	78.77	0	28	189.62	0	32

	Single-Glazed, Southwest			Double-Glazed, Southwest		
	$q_{s,opt}$	ψ_{opt}	Σ_{opt}	$q_{s,opt}$	ψ_{opt}	Σ_{opt}
Kerman	120.87	48	37	282.85	53	43
Ahwaz	190.37	51	41	327.03	53	44
Bandar Anzali	93.44	45	39	255.95	49	46
Chabahar	173.21	53	37	325.34	57	41
Kamyaran	118.24	49	39	275.99	51	45

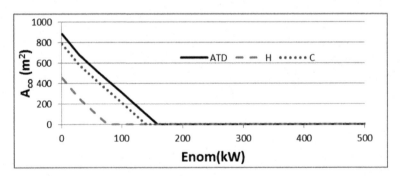

Figure 7.6 Collector size versus engine size for intg-DHW, Kamyaran.

building. In addition the auxiliary boiler size is depicted in Figure 7.7 and it shows that the boiler size in different strategies is approximately the same and similar to that calculated for the basic CCHP system. The reason is that the solar system cannot receive thermal energy at night, therefore the system should be designed to provide all the energy demands at all times. It is worth mentioning that using a thermal storage system decreases the boiler size because it can store surplus heat during daytime and reuse it at night.

To determine the engine size and collector area, the fitness function is utilized, and the results for the five climates are presented in Table 7.2.

According to Table 7.2, the ff recommends not using a solar collector in most cases. After determining the economic criteria, it was revealed that the solar collector price is still too high and cannot compete with other equipment such as the heat recovery of

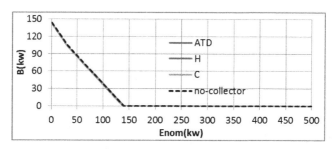

Figure 7.7 Boiler size versus engine size for basic and hybrid CCHP, Kamyaran.

Table 7.2 **Engine Size and Collector Area for Different Strategies in Five Climates**

Climate		ATD	H	C
Ahwaz	$E_{nom}(kW)$	33	288	33
	$A_{co}(m^2)$	1451	0	1459
Kerman	$E_{nom}(kW)$	118	58	100
	$A_{co}(m^2)$	0.43	0	0
Bandar Anzali	$E_{nom}(kW)$	29	117	155
	$A_{co}(m^2)$	788	0	0
Chabahar	$E_{nom}(kW)$	255	159	236
	$A_{co}(m^2)$	0	0	0
Kamyaran	$E_{nom}(kW)$	160	78	141
	$A_{co}(m^2)$	0	0	0

an engine or auxiliary boiler. In some cases such as Ahwaz or Bandar Anzali, the recommended size of the collector has increased dramatically on one side and the engine size has decreased significantly on the other side. Using 1451 m² of solar collector (for Ahwaz) on the roof of the building that has been analyzed is impossible. In order to make solar collectors more popular it is recommended that they be supported by governments, banks, and energy and environmental organizations through subsidies, proper loans, carbon credits, etc.

7.5 Problems

1. According to the formulations presented for the plate collector, determine the optimum direction for a solar collector in the city where you live.
2. If you have written code for the problem 6 presented in the previous chapter, couple it with the solar code written for Problem 1.
3. The hybrid CCHP in this chapter is designed based on *ff*; use the MCSF to redesign the hybrid CCHP system and compare the results with the *ff*.

4. If you are required to use a concentric parabolic collector (CPC), what would be the difference in the formulation in comparison to the plate collector? You can use [10, 14] to find the difference.
5. What would be the advantages of using PVT technology instead of a plate collector with the CCHP designed in this chapter? Does it reduce the engine size? What would be your prediction about the auxiliary boiler size?
6. Discuss the technological, economic, and environmental characteristics of using a thermal storage system instead of the auxiliary boiler when the need arises.
7. Assume we have a CCHP system integrated with PVT technology and a combination of electrical and absorption chillers. The absorption chiller is supposed to be used for the cooling base load and the electrical chiller is operated at cooling peak loads. Draw the schematic of this CCHP system and modify the formulations for this novel CCHP system.

References

[1] Oliveira, A.C., Afonso, C., Matos, J., Riat, S., Nguyen, M., Doherty, P., 2002. A Combined Heat and Power System for Buildings Driven by Solar Energy and Gas. Applied Thermal Engineering 22, 587–593.

[2] Yagoub, W., Doherty, P., Riffat, S.B., 2006. Solar Energy-Gas Driven Micro-CHP System for an Office Building. Applied Thermal Engineering 26, 1604–1610.

[3] Wang, J., Dai, Y., Gao, L., Ma, S., 2009. A New Combined Cooling, Heating and Power System Driven by Solar Energy. Renewable Energy 34, 2780–2788.

[4] Fraisse, G., Bai, Y., Le Pierre'S, N., Letz, T., 2009. Comparative Study of Various Optimization Criteria for SDHWS and a Suggestion for a New Global Evaluation. Solar Energy 83, 232–245.

[5] Luo, H., Wangb, R., Dai, Y., 2010. The Effects of Operation Parameter on the Performance of a Solar-Powered Adsorption Chiller. Applied Energy 87, 3018–3022.

[6] Parida, B., Iniyan, S., Goic, R., 2011. A Review of Solar Photovoltaic Technologies. Renewable and Sustainable Energy Reviews 15, 1625–1636.

[7] Hassan, H.Z., Mohamad, A.A., Bennacer, R., 2011. Simulation of an Adsorption Solar Cooling System. Energy 36, 530–537.

[8] Wang, J., Zhao, P., Niu, X., Dai, Y., 2012. Parametric Analysis of a New Combined Cooling, Heating and Power System with Transcritical CO_2 Driven by Solar Energy. Applied Energy 94, 58–64.

[9] Conserval Engineering Inc. www.SolarWall.com.

[10] ASHRAE Handbook 2007 - HVAC Applications SI, Chapter 33, Solar Energy Use.

[11] ASHRAE Handbook 2008 - HVAC Systems and Equipment IP, Chapter 36, Solar Energy Equipment.

[12] WOLF company,Germany, Änderungen vorbehalten, http://www.wolf-heiztechnik.de/ 11-Feb-2014.

[13] Chromasun, Inc. USA http://www.chromasun.com/11-Feb-2014.

[14] Sukhatme, S.P., 1984. Solar Energy: Principles of Thermal Collection and Storage. India: Tata McGraw-Hill Publishing Company Limited, New Delhi.

[15] Şen, Z., 2008. Solar Energy Fundamentals and Modeling Techniques. Springer-Verlag London Limited.

CCHP Thermal Energy Storage

8.1 Introduction

Different forms of energy can be stored for use later. For example, pump storage
power plants use electricity to pump and store water in high altitude storage during
hours of low demand for electricity and use the water to produce electricity by running it over turbines during peak hours of demand. Another example of mechanical
energy storage are compressed air energy storage systems in which the compressed air
is stored during times of low demand used again during peak times. Electricity storage systems or batteries are connected to direct electric current to be charged during
low load periods. At the time of discharge the chemical energy is released to produce
electricity during periods of high load. Energy can also be stored in the form of heat or
cold in thermal storage materials such as water, rocks, pebbles, phase-change materials, etc. For example solar air heating systems can use pebbles to store extra collected
solar heat and use it when the need arises.

8.2 Thermal Energy Storage (TES)

Thermal storage systems remove heat from or add heat to a storage medium for use at
another time [1]. Cold, like heat, also can be stored in TES systems. TES systems are
commonly used with solar heat collectors to store the collected solar heat. They also
can be used when there is a potential for heat loss in a process or piece of equipment
and there is the possibility of using this heat at another time for process or heating
purposes. TES systems can be used to store extra cold produced by a chiller during
periods of low cooling demand for reuse during peak hours.

   TES systems can be very helpful in improving the performance of energy conversion equipment such as CCHP cycles. In the CCHP systems designed up to now in
this book, the surplus electricity is sold back to the grid, but the excess heat is wasted
to the ambient. A TES system stores the surplus heat, which is usually wasted, and
reuses it whenever needed. This ability improves the reliability of the CCHP system
as well as the energy utilization factor. Using a well-designed TES system can omit
or downsize the auxiliary boiler in the CCHP system, which means cost reduction,
fuel savings, and pollution reduction. It also provides backup capacity for the consumer, increases the flexibility of the CCHP cycle, and extends the capacity of the
basic CCHP cycle. Energy can be stored in different liquids, solids, or phase-change
materials (PCMs). The thermal storage medium should be inexpensive, easily available, environmentally friendly, nonflammable, nontoxic, nonexplosive, noncorrosive,
neutral, inert, and compatible with the CCHP system. The medium should have good
physical properties such as high density, high sensible and latent heat properties, good

heat transfer properties such as conductivity, and stable properties in thermal cycling conditions. Water is the most commonly used material for TES systems since it is easily available, inexpensive, nontoxic, neutral, nonflammable, nonexplosive, and has an acceptable range of working temperatures in the liquid form at atmospheric pressure. Furthermore water TES systems are compact (roughly one-third of a pebble bed) and compatible with hydronic heating systems. Besides these advantages, water treatment should be done to prevent scaling and corrosion, especially if it is used for heat storage. Scaling and corrosion are not as severe in the cold storage as they are in heat storage systems.

The most commonly used solid material for TES in air systems is rocks or regenerated matrix made from concrete masonry units (CMUs); in addition gravel is popular in the air systems because it is inexpensive and plentiful [1,2].

Ref. [1] proposes using TES when one or more of the following conditions exist:

When loads are of short duration, loads occur infrequently, loads are cyclical, loads are not coincident with the available energy source, energy costs are time dependant, charges for peak power demand are high, utility rebates, tax credits or other economic incentives are provided for using load-shifting equipment, energy supply is limited thus limiting or preventing the use of full-size non-storage systems, facility expansion is planned, and the existing heating or cooling equipment is insufficient to meet the new peak load but has spare nonpeak capacity or interruption in cooling water cannot be tolerated by a mission-critical operation.

TES systems can be classified into sensible heat storage (SHS) and latent heat storage (LHS).

8.2.1 Sensible Heat Storage (SHS)

In SHS systems no phase change occurs and only the temperature of the storage medium is increased or decreased. Heat can be stored by raising the temperature of the storage medium while cold is stored by lowering the temperature of the storage medium. The amount of heat that is stored and can be used in the SHS system depends on the low and high temperature limits (T_l, T_h) of the SHS system. Q_{TES} (storage capacity) for heat can be determined according to the following equation:

$$Q_{TES}\big|_{T_l}^{T_h} = \int_{T_l}^{T_h} mC_p\, dT = m\overline{C}_p\Delta T, \Delta T = T_h - T_l \tag{8-1}$$

where m, C_p, and $\overline{C}_p$ are the mass of the SHS medium, specific heat capacity, and average specific heat capacity of the SHS medium, respectively. The average of C_p is based on T_l and T_h. In the liquid SHS medium, the mass of liquid medium can be related to the volume of the storage as follows:

$$Q_{TES}\big|_{T_l}^{T_h} = \rho(V_{strg} - V_{ect})\overline{C}_p(T_h - T_l) \approx \rho(V_{strg})\overline{C}_p\Delta T \tag{8-2}$$

Table 8.1 **Some Solid and Liquid Mediums Used in SHS Systems [3]**

Medium	Medium Type	Temperature Range (°C)	Density (kg/m³)	Specific Heat (J/kg.K)
Rock	Solid	20	2560	879
Brick	Solid	20	1600	840
Concrete	Solid	20	1900–2300	880
Water	Liquid water	0–100	1000	4190
Caloria HT43	Oil	12–260	867	2200
Engine oil	Oil	Up to 160	888	1880
Ethanol	Organic	Up to 78	790	2400
Proponal	Organic	Up to 97	800	2500
Butanol	Organic	Up to 118	809	2400
Isotunaol	Organic	Up to 100	808	3000
Isopentanol	Organic	Up to 148	831	2200
Octane	Organic	Up to 126	704	2400

where ρ, V_{strg}, and V_{ect} are the medium density, total storage volume, and volume occupied by the coils and other equipment inside the storage, respectively.

The properties of some solid and liquid mediums used in SHS systems are listed in Table 8.1.

8.2.2 Latent Heat Storage (LHS)

LHS systems achieve most of their heat storage capacity from the latent heat of phase change of materials, which can be used as the TES medium. This phase change includes solid to liquid, liquid to gas, and vice versa. LHS mediums can be classified as organic (paraffin and nonparaffin), inorganic (salt hydrate and metallic), and eutectic. For comprehensive information about TES mediums, especially PCMs, the readers are advised to read Ref. [3]. The heat stored in a PCM during heat absorption can be calculated as follows:

$$Q_{TES}\big|_{T_l}^{T_h} = \underbrace{\int_{T_l}^{T_m} mC_p\, dT}_{SHS\ in\ solid\ form} + \underbrace{ma_m \Delta h_m}_{LHS\ while\ melting} + \underbrace{\int_{T_m}^{T_h} mC_p\, dT}_{SHS\ in\ liquid\ form}$$

$$Q_{TES}\big|_{T_l}^{T_h} = m[\overline{C}_{sp}(T_m - T_l) + a_m \Delta h_m + \overline{C}_{lp}(T_h - T_m)]$$

(8-3)

where a_m is the fraction of PCM melted, and $\overline{C}_{sp}$ and $\overline{C}_{lp}$ are the average specific heat of solid and liquid PCM in the temperature rages of $T_m - T_l$ and $T_h - T_m$, respectively. It is evident that while $a_m < 1$ the third term (SHS in liquid form) is still zero because it is assumed that melting occurs at constant temperature and this continues until all the solid melts.

8.3   Charge and Discharge of TES

A CCHP system that is equipped with a TES system can benefit from the change-able demands of the consumer. In basic CCHP systems, when the recoverable heat is higher than the ATD, the surplus heat is wasted. If the CCHP system was able to store the extra recoverable heat it could be used later when the demand becomes higher than the recoverable heat. It is evident that using a well-designed TES system for a CCHP system results in more fuel saving and pollution reduction. It can also improve the economic benefits of the CCHP system since the prime mover size or the auxiliary boiler can be smaller. A basic CCHP system that is equipped with a solar collector and TES creates even more fuel saving and pollution reduction.

In general the TES will be charged when

$$Q_{rs} = Q_{solar} + Q_{rec} > ATD \ \& \ T_{TES} < T_h \tag{8-4}$$

Technical limitations may put a constraint as the maximum operating temperature (T_h) of the TES. For example if we use water as the TES medium in an atmospheric pressure storage, the addition of heat to water when its temperature reaches its satura-tion temperature (100°C) causes the water to start evaporating, which may increase the pressure inside the storage and cause damage to the storage and instrumentation. In addition, corrosion increases at higher temperatures. Therefore in this case $T_h = 100°C$. The amount of heat charge can be calculated as follows:

$$Q_{charge} = \begin{cases} \varepsilon_{TES}(Q_{rs} - ATD) & if \ (T_{TES} < T_h \wedge ATD < Q_{rs}) \\ Q_{TES} & if \ (T_{TES} = T_h \wedge ATD < Q_{rs}) \end{cases} \tag{8-5}$$

where Q_{TES} is the TES capacity, and ε_{TES} is the effectiveness of the storage in receiv-ing heat.

The TES will be discharged when

$$Q_{rs} = Q_{solar} + Q_{rec} < ATD \ \& \ T_{TES} > T_l \tag{8-6}$$

It should be noted that according to the technical limitations, a minimum tempera-ture (T_l) may be designed for the TES, in which discharging is not possible when the TES temperature falls below T_l. This minimum temperature may depend on the work-ing fluid output temperature from the solar and recovery systems and the storage and its instrumentation.

The amount of dischargeable heat from the TES system depends on the instantane-ous temperature of the TES medium (T_{TES}), and can be calculated as follows:

$$Q_{discharge} = \begin{cases} \varepsilon_{TES} \ Q_{TES} \big|_{T_l}^{T_{TES}} & if \ (T_{TES} > T_l \wedge ATD > Q_{rs}) \\ 0 & if \ (T_{TES} = T_l \vee ATD < Q_{rs}) \end{cases} \tag{8-7}$$

8.4 Sizing of TES

Proper sizing of TES, like other components of the CCHP system, requires accurate calculation of consumer demands and load profiles. An undersized TES fails to cover all energy demands when the load exceeds the capacity of the TES system. On the contrary an oversized TES reduces its economic benefits. This means that to best design a TES system for integration in the CCHP system, thermodynamic, environmental, and economic criteria should be evaluated simultaneously. In addition to the load profiles, the operational strategy of the TES also has a significant impact on the design criteria. Since cooling and heating loads are very changeable in different climates, the TES capacity can be designed to cover the heating, cooling, DHW, or ATD of the consumer in conjunction with the recoverable heat from the prime mover and solar collector.

ATD-strategy: If the TES is designed to cover the annual maximum ATD with assistance from the heat recovery and solar systems (figure 8.1), its size can be related to the solar collector and prime mover size according to energy balance as follows:

$$Q_{TES}\big|_{T_l}^{T_h} = ATD_{\max} - Q_{rec,\min} - Q_{solar}, \; ATD_{\max} \geq Q_{rec,\min} + Q_{solar} \qquad (8\text{-}8)$$

where

$$Q_{rec,\min} = f(E_{nom})$$
$$Q_{solar} = A_{co} \cdot q_{s,opt} \qquad (8\text{-}9)$$

In the CCHP system and the optimization process, the prime mover size is the first priority. Therefore the engine size should be adjusted based on the above equation and the solar and TES systems should also be changed to satisfy the energy balance equation.

H-strategy: If the TES is designed to provide the maximum heating load demand ($H_{dem,\,\max}$) with cooperation from the heat recovery system and solar collector, we have

$$Q_{TES}\big|_{T_l}^{T_h} = H_{dem,\max} - Q_{rec,\min} - Q_{solar}, \; H_{dem,\max} \geq Q_{rec,\min} + Q_{solar} \qquad (8\text{-}10)$$

C-strategy: If the TES is designed to fulfill the maximum cooling load ($C_{dem,\,\max}$), we get

$$Q_{TES}\big|_{T_l}^{T_h} = \frac{C_{dem,\max} - COP_{abc} \cdot (Q_{rec,\min} + Q_{solar})}{COP_{abc}},$$
$$C_{dem,\max} \geq COP_{abc} \cdot (Q_{rec,\min} + Q_{solar}) \qquad (8\text{-}11)$$

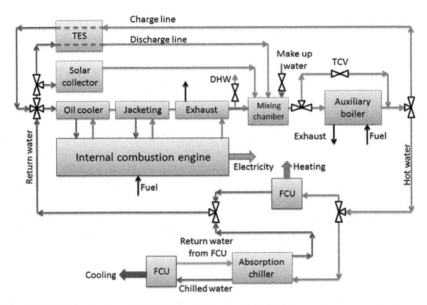

Figure 8.1 Solar collector and TES integrated in the CCHP cycle.

D-strategy: If the TES designed to provide $D_{dem,\,max}$, in conjunction with the solar collector, we have

$$Q_{TES}\big|_{T_l}^{T_h} = D_{dem,\,max} - Q_{solar}, \; D_{dem,\,max} \geq Q_{solar} \tag{8-12}$$

If DHW is integrated in the heat load of the consumer (*intg-DHW*), the D-strategy no longer needs to be considered. But if the DHW load is not included in H_{dem} (*sprt-DHW*) the ATD-strategy will be omitted from the analyses.

An interesting case is when water is chosen as the TES medium. In this case, since the density, average specific capacity, and operation temperature range of water are constant, the storage volume can be determined as follows:

$$V_{strg} \approx \frac{Q_{TES}\big|_{T_l}^{T_h}}{\rho \overline{C}_p \Delta T} \tag{8-13}$$

References

[1] ASHRAE Handbook 2007 - HVAC Applications SI, Chapter 34, Thermal Storage.
[2] ASHRAE Handbook 2008 - HVAC Systems and Equipment IP, Chapter 36, Solar Energy Equipment.
[3] Sharma, A., Tyagi, V.V., Chen, C.R., Buddhi, D., 2009. Review on Thermal Energy Storage with Phase Change Materials and Applications. Renewable and Sustainable Energy Reviews 13, 318–345.

CCHP Operation and Maintenance

9

9.1 Introduction

A well-designed CCHP system will fail to operate efficiently if an accurate and effective operation and maintenance (O&M) program is not applied. An improper O&M program shortens the lifetime of equipment, decreases cycle efficiency, increases fuel consumption and environmental pollution, and reduces the economic benefits of the CCHP system. On the contrary, applying a proper operation and maintenance program to the CCHP system and its components extends the project lifetime, and maintains the cycle's optimum efficiency, pollution reduction rate, and economic benefits.

In this chapter a general O&M method will be introduced but the details of operation and maintenance differs from one type of equipment to another. In addition different manufacturers propose different O&M programs for their products. Therefore the best way to find the details on O&M is to study the manufacturer's documents.

9.2 General O&M Program

A well-designed O&M program includes correct pre-commissioning, commissioning, post-commissioning, and shutdown. In addition, filling out the log sheets correctly and analyzing them along with precise troubleshooting are vital for a successful O&M program.

9.2.1 Pre-commissioning

A pre-commissioning program includes but is not limited to the following tasks that the operator should check them before startup of the CCHP system. It is recommended to use the operation manuals provided by the manufacturers, as these manuals are the main source of O&M information. A more professional method is to provide a checklist of actions that should be completed before startup and follow the checklist for each pre-commissioning of the CCHP system.

1. Always follow the safety procedures and use safety equipment such as proper boots, clothing, gloves, goggles, etc.
2. Make sure that no tag-out/lockout is applied to equipment such as motors, valves, etc.
3. Make sure that all of the electrical connections are safe and correct (check for correct grounding, check electrical insulation for uncovered wires, connection to the grid, correct direction of rotation of the electrical motors used for driving pumps, fans, and other rotary equipment, etc.).
4. Check the lubrication system (check the oil level in the oil reservoir, look for oil leakage in the oil piping, connections, and components that are supposed to be lubricated, make sure that the oil recommended by the manufacturer is used, make sure that the oil filters are not

clogged, make sure that the oil coolers can operate properly, if oil is water cooled make sure water is there and look for water leakage and if it is air cooled make sure that fans are installed correctly, if some components are lubricated by grease check that the grease meets standards and is sufficient and clean).

5. Make sure that all the mechanical connections are well connected. (look for loose-fitting, misaligned couplings, belts, and pipe connections, avoid incorrect positioning of valves, etc.)

6. Check the fuel, the fuel pressure, and temperature requirements.

7. Check the sealing system and its auxiliary subsystems (pump flushing line, gas sealing lines of compressor and gas turbine, oil seal line, steam sealing line, etc.).

8. Check the cooling systems (jacketing of engines, motors, bearings, lube oil cooling, etc.).

9. Check all of the filters and strainers for clogging or failure (air intake filters of gas turbines and compressors, lubrication system filters, suction pump strainer, fuel filters, etc.).

10. Look for closed inlet air baffles that provide enough fresh air for combustion in the prime mover, boiler, etc.

11. Make sure that all of the measurement instrumentation is calibrated and that the calibration date has not expired.

12. Check the condition of thermal insulation used for hot or cold lines, TES systems, etc. Never press the insulation; this reduces the thickness and as a result the thermal resistance decreases and energy loss increases.

13. If the CCHP system has a liquid-type solar collector, make sure it is protected from freezing in a cold climate. Also make sure it is placed in the optimum direction to receive maximum solar heat.

14. Check the hot and cold connections to heat exchangers and make sure they are not connected incorrectly. In the shell and tube exchangers water flows on the tube side most of the time, because fouling cleaning on the tube side is much easier than on the shell side.

9.2.2 Commissioning

After pre-commissioning of the CCHP system, startup is the next step. The commissioning in general includes but is not limited to the following steps. Make sure that the pre-commissioning steps are completed correctly and completely.

1. Inform your supervisor and other colleagues about the commissioning.

2. Run the lubrication system correctly. The lubrication system is the first system that should be run during startup because most of the component wear happens due to incorrect startup. The lubrication system usually includes a reservoir, oil heater, strainer, main and auxiliary pump, recirculation pump, main and auxiliary filter, main and auxiliary cooler, pressure gages, thermometer level controls, etc. Check the oil temperature; usually the best temperature for lubrication oil is about 40°C. If the oil temperature is low, start the oil heater and circulation pump to increase the oil temperature homogeneously. Prime the auxiliary components of the lubrication system by starting the auxiliary pump. Priming includes venting, filling with oil, and making the components ready for use in standby. After priming, follow the manufacturer recommended steps to run the main pump to start lubrication. If the lube oil pump is a positive displacement type (reciprocating, gear, screw, etc.) make sure that the discharge valve is open to avoid extreme pressure that may cause damage to the components. If the pump is centrifugal, the discharge valve should be closed at the startup and opened slowly and gradually; in this step the min flow line should stay open until the discharge valve is fully open.

3. Bring the cooling systems into operation. The cooling systems may include the air cooler or cooling tower and heat exchangers. The cooling towers have a fan and pump, but air coolers usually have two fans. This rotary equipment should be run according to the manufacturer guidelines. Always check the water quality, scale formation in the heat exchangers, and the cooling tower's fill packing.

4. Bring the sealing systems into operation. Running the prime mover, pumps, and other rotary equipment before running the sealing systems can cause serious damage to the sealing systems and other components. Make sure that enough flow with proper pressure will be used in the flushing and sealing systems.

5. Bring the heat recovery heat exchangers into operation. In the heat exchangers the cold fluid should flow first into the heat exchanger and then the hot fluid. Hot and cold lines should be opened gradually to avoid thermal stress in the exchanger.

6. If cooling is required, the auxiliary systems of the cooling system should be operated. For absorption chillers the refrigerant and solution pumps, vacuum pump, cooling tower pumps, cooling water pump, chilled water pump, etc. should be operated. At this stage the auxiliary systems use grid electricity for operation. When the prime mover is operated, the auxiliary systems can use the CCHP electricity and become disconnected from the grid. The compression chillers use electricity, therefore they can stay offline until the CCHP can provide enough electricity to use the cooling system. It is very important to follow the manufacturer instructions for the startup of the cooling systems.

7. Run the prime mover. The startup procedure of prime mover very much depends on the prime mover type. The procedure recommended by the manufacturer must be followed to run the prime mover safely and correctly.

8. The auxiliary boiler starts working automatically whenever needed. But it should be ready for operation.

9. If the CCHP system has a solar collector, the solar systems enter into operation to use solar heat when the sun is shining.

10. If the CCHP system has a TES system, bring it online so it can charge.

11. If the CCHP system uses electricity storage systems, bring them online to store surplus electricity for selling or using during electrical peak hours.

9.2.3   Post-commissioning

After startup or commissioning of the CCHP system, the operator still has very important duties. Some of these tasks should be done just after the commissioning and others should be done at certain times when the CCHP system is operating. The following procedure can be considered a guideline, but the operators should be aware that these guidelines are general; they may need to revise and extend them for their own CCHP system and its components.

General routine checks in the first hour after commissioning are recommended as below; the operator may need to check some of these items several times in the first hour of operation to confirm the safe operation of the system.

1. Make sure that all the commissioning steps are taken correctly and completely. It is more convenient to provide a checklist for this purpose and tick the commissioning procedure step by step.

2. Listen for cavitation noise in the pumps. It sounds like small sand particles circulating inside the pump.

3. Check the bearings temperature. An unusually hot bearing is a sign of bearing failure, wrong oil type, oil pollution, oil corruption, shaft misalignment, excessive vibration, shaft run out, mass unbalance, etc.
4. Check for excessive vibration in the bearings, casings, and couplings. Vibration may be due to misalignment, bearing damage, shaft run out, cavitation, mass unbalance, etc.
5. Check for sudden shaking in the piping, fittings, and valves that may be a sign of water hammering.
6. Check for any type of leakage in the heat exchangers, lube oil systems, hot and cold pipelines, etc. Check the permissible leakage of packing-type pump sealing systems. Packing-type sealing systems need cooling. If water is used for to cool the packing, 60 droplets of water per minute is required for proper cooling. Less leakage results in packing overheating and burning and finally excessive leakage. Leakage of more than 60 droplets per minute may be due to a loose gland, packing failure, and incorrect placement of packing rings and lantern rings in the stuffing box.
7. Check that the amperage of electrical motors is in range. High amperage is a sign of improper operation or overloading of rotary equipment connected to the motor.
8. Check the pressure gages for fluctuating pressure at the pump discharges. Fluctuating pressure may be due to cavitation, impeller failure, etc.
9. Check the pumps for fluctuating flow rate. Fluctuation in the pump flow rate may be due to cavitation, impeller failure, low level fluid in the fluid source (cooling tower basin, oil reservoir, etc.).
10. Check the oil temperature and pressure before entering the bearings.
11. Check that the cooling water flow and temperature is in the correct range.
12. Check the differential pressure of the filters and strainers.
13. Check the oil temperature and pressure before and after lubrication.
14. Check the differential pressure (DP) of heat exchangers; high DP is a sign of excessive scale formation and fouling in the heat exchangers.
15. Check that the min flow line of pumps is closed to avoid wasting of energy.
16. Check that all the measurement instruments are working and none of them is stuck or stopped.
17. Check all of the fans, and their coupling and motors. Listen for unusual noise.
18. Check the cooling tower for excessive drift and for any nozzles that may not be working; also check the PH and hardness of water.
19. Check the electrical, heating, and cooling terminals of the consumer and make sure they receive the required energy type with the best quality.
20. Make sure that you have checked all the tasks given to you in the checklists for equipment. When you are certain about pre-commissioning, commissioning, and post-commissioning procedures you can plan your next visit to the CCHP system equipment to check the operating log sheets.

9.2.4 Operating Log Sheets

After completing the pre-commissioning, commissioning, and post-commissioning checklists a successful operator follows up on the operation of the CCHP cycle to avoid the development of technical problems. Recognizing a problem when it starts to show up lets us fix the problem before it develops and become a bigger problem that may cause an unscheduled shutdown or emergency.

Operation log sheets contain some parameters that should be read or checked by the operators from the control system or measurement instruments. Usually a permissible variation range is defined for parameters such as pressure, temperature, vibration, etc. Continuous records of these parameters can reveal the trend of parameters (i.e., if they are steadily increasing, steadily decreasing, fluctuating, or steadily constant). These trends help us to recognize what parameters are going to make trouble in the CCHP system and its components. Therefore, in addition to reading the correct data from the CCHP components, analyzing the log sheets is the most important task of the operator and supervisor in recognizing the problems before development of potential cause emergencies.

Here three simple but different examples are presented to show that detecting them as they start gives us the opportunity to repair them at the right time without paying extra costs or causing an unscheduled shutdown.

Example 1: The suction strainer of a pump is clogged.

A DP increase through the pump's suction strainer can be solved easily by cleaning the strainer; this is what a successful operator does. However, if the pressure drop across the suction strainer increases further it can result in low-pressure cavitation in the pump suction eye. Cavitation erodes the internal parts of the pump; generates noise and vibration; makes the discharge pressure and flow fluctuate; increases the axial and radial forces, thus overheating and damaging bearings; increases the oil temperature; and also creates problems for the sealing system. As can be seen timely recognition of the problem and repairing it promptly avoids all the negative consequences.

Example 2: The water level in the cooling tower is decreasing.

Decreasing water level in the basin of a cooling tower can be simply recognized using the recorded data on the log sheets of the cooling tower. A successful operator is one who undertakes prompt action to increase the water level to avoid the hazardous effects of low water level, and at the same time starts a troubleshooting procedure to find the main reason for water loss in the system to remove the problem completely.

It should be mentioned that accurate troubleshooting avoids repetition of the same trouble in the CCHP cycle. For example, prompt action to increase the water level in the cooling tower in this example is an introductory action to avoid further damage, but if we do not find the main reason for the water loss, the water consumption of the cooling tower increases steadily, which is unfavorable. A favorable action is to remove the main reason for water loss in the system, and at the same time doing prompt actions to avoid damage in the CCHP components due to low water level in the cooling tower.

Example 3: Water flow of heat recovery systems at constant load is increasing.

Increasing water flow rate of a heat recovery system can be easily recognized according to the recordings in the log sheets. Extra water flow means higher pump energy consumption. Over the long term, the pumping cost increases significantly. In addition, in the design step, pumps are selected based on the best efficiency point (BEP) operation. This means that the pump is chosen to operate as close as possible to the operating point with the highest efficiency. When the flow rate has changed significantly, the operation point would be displaced and moves away from the pump BEP. In this condition the loss of pumping energy due to low efficiency operation of the pump will be added to the costs of a high flow rate in the heat recovery system.

Operators always should look for the easy explanations first. This means the reasons that do not require any shut down or disassembling of equipment from the CCHP system.

Increased flow rate may be due to leakage in the pump sealing, piping, or exchangers; installation of new consumer terminals such as fan coils; uncalibrated flow measurement instruments; increased heat loss energy from the prime mover due to increasing water jacketing, oil cooler, or exhaust temperatures, etc.

As can be seen, an increase in the water flow rate for heat recovery can occur for a span of reasons that ranges from easy to difficult. The troubleshooter and operator should categorize the reasons from simple to difficult and investigate the solutions in that order. If solving a problem requires shutdown or disassembling equipment it should be done only with written permission from the supervisor, owner, and consumers of the CCHP system. To investigate the trouble and conduct the troubleshooting in as effective a manner as possible, a guideline is presented below.

9.2.5 Troubleshooting

Successful troubleshooters can save significant maintenance and part replacement costs. They analyze the log sheet data and operator reports to find trouble points. They find the main reason for problems and propose possible solutions while avoiding any extra maintenance jobs that may result in extra cost and cause other problems for the components.

A general procedure for troubleshooting of equipment is proposed in the following:

1. First of all define the problem clearly. What exactly is the problem? What is the expected situation? What is the current condition? And what was the previous situation?
2. Find out about the history of the equipment. How long it has been working? When was the last overhaul of the CCHP system and its components? What changes have been made to the CCHP system and its components? In comparison with the previous overhauls were there repetitive changes and part replacement? When did the current problem start? Was it just after overhaul or after a long period of operation after the overhaul? How has the problem developed? Was it quick or gradual? Look for trends of different parameters, such as temperature, pressure, flow rate, vibration, etc. Are these parameters constant, steadily increasing, steadily decreasing, or fluctuating? Is the problem increasing, decreasing, fluctuating, or constant?
3. Verification. Make sure that all of the readings are correct. Make sure that the calibration date has not expired. Make sure that no measurement instrument is broken or stuck or has stopped working.
4. Requirements for troubleshooting. Everything that may assist you for troubleshooting should be available including operation log sheets; pre-commissioning, commissioning, and post-commissioning checklists; P&I diagrams; cross-section drawings of equipment; operation instructions from the manufacturer; operator reports; reports of previous overhaul and maintenance jobs; and last recoded data from the equipment.
5. Make a list of reasons that may cause the current problem in the CCHP system and categorize them from simple to hard according to cost, maintenance job, shutdown requirement, equipment disassembly, etc.
6. Apply for maintenance after you have informed your supervisor, CCHP owner, and consumers.

As a sample, in the following some general troubles that may occur in the equipment are analyzed.

Example 4: The lubrication oil temperature has increased.

Possible reasons, corrective actions, and recognition method:

1. Incorrect oil type is possibly used; check the certificates of the recommended oil and compare it with the newly used oil.

2. Oil is foaming excessively due to losing its anti-foam additives, or water and air leakage into the oil system. Oil analysis can be done to find out about it. Water leakage into the oil can be tested by getting sample oil from the drain valve of the oil reservoir. Excessive foam can be pumped into the lubrication system. As a result not enough oil will be used for lubrication and oil temperature increases rapidly. Excessive foam can even create fire in hot areas especially those in contact with some flammable insulation.

3. Oil is polluted due to oxidation, ambient pollution such dust particles, etc. This pollution decreases the lubrication properties of oil and the oil temperature increases accordingly. Oil analyses can reveal the pollution type.

4. If the oil cooler is a water-type exchanger, it may have been fouled and the cooling process may not have been conducted properly. This can be recognized by looking at the oil temperature difference variation and water pressure drop change across the oil cooler. In this case the power consumption of the electromotor increases as well.

5. If there is a water-type oil cooler, the water temperature has possibly increased due to incorrect operation of the cooling tower. If this is the reason, the cooling tower must be analyzed to find the main reason for its incorrect operation. For example, it may be due to loss of pump capacity, excessive drift, insufficient make-up water, fouling in the fill packing, improper distribution of water by the tower's nozzles, etc. It should be noted that if the loss of pump capacity is the main reason, it should be analyzed as well

6. If there is an air-cooled oil cooler, the fans are probably off or not working properly. Visit the fans and check to find out.

7. If an air-cooled oil cooler is used, the elevated ambient air temperature can decrease the cooler efficiency and increase the oil temperature. Check the ambient temperature. If it is high, you can increase the fan speed or use water spray between the exchanger and fans as auxiliary evaporative cooling during hot times of the day.

8. Failure of bearings. Failure of a bearing also creates noise and vibration. Therefore it can be recognized based on the noise or vibration. If a bearing has failed, corrective actions must be taken as soon as possible because it can produce serious problems for other parts as well. For example the vibration due to bearing damage may cause damage to the sealing system with very narrow clearances, hazardous vibration for rigid couplings, wearing of labyrinth and wearing rings, etc.

9. Misalignment in the couplings or bearings. Misalignment generates vibration in the bearings and couplings as well.

10. Oil filters have clogged. Due to clogging the filters, less oil flows to the bearings and as a result the oil becomes hotter. This can be recognized according the pressure drop across the filters.

11. The oil heater inside the reservoir is working while it should be off. Check if the heater is on or off.

12. Thermometer is out of calibration. Check the oil temperature with a portable thermometer in the oil reservoir.

13. Other possible causes.

After listing the possible reasons, classify them into 3 classes of "only small change or adjustment is required, no shutdown, no disassembling," "shutdown is required, no disassembling is needed," and "shutdown and disassembling are required" and check the reasons from the first class to the third class.

Example 5: Chilled water temperature from the absorption chiller to the fan coils has increased.

1. Heat source energy is insufficient. If the chiller is indirect fire, the heat source temperature, pressure, or its flow rate has probably decreased and vaporization of the refrigerant in the generator has decreased accordingly. Check the flow rate, temperature, and pressure of the heat source entering the generator.
2. Heat source is not sufficient. If the chiller is direct fire there may be insufficient fuel or oxygen for completing the combustion process. Check the inlet air baffle positions and fuel pressure.
3. Vacuum inside the evaporator is broken. The vacuum pump has probably failed or the chiller has air leaking from ambient to the chiller. Check the vacuum pump operation, and use leakage discovery methods.
4. The temperature of return water from the cooling tower is high. High temperature water from the cooling tower is a problem for the condenser and absorber of the chiller. This can be easily checked by using thermometers to check the return water temperature. This problem can occur due to scale formation on the fill packing or in the spray nozzle, improper working of the tower fan, etc.
5. The flow rate of water to the condenser is low. To find out about this problem you can check the pump discharge valve position, check for cavitation, check the water level in the tower basin, and check the pump suction strainer for clogging,
6. Cooling load has increased. This can be checked by looking at the outdoor temperature and return water temperature from the terminal consumers. This may happen due to newly installed terminal consumers, increase in outdoor temperature, etc.
7. Cold line insulation have been removed, cut, scratched, or compressed. Check the insulation for any deformation or removal.
8. The thermometer is not calibrated. Check the calibration of the thermometer.
9. Refrigerant pump is leaking and the refrigerant level in the chiller has decreased. Check the refrigerant pump for leakage from the sealing system, connections, and flanges.
10. The condenser tubes are fouled from the inside. Check the amperage of pump's electromotor. Increase in the electromotor amperes is a sign of increasing friction head loss of the pump. In addition you can check the hardness and PH of water in the cooling tower. PH higher than 7.5 can increase scale formation rapidly, while PH smaller than 6.5 increases chemical corrosion.

More information about the operation and maintenance of CCHP equipment can be found in the manufacturer documents and [1–5].

References

[1] Fletcher, J., 1999. HVAC Troubleshooting Manual. The Building Services Research and Information Association (BSRIA). Printed by The Chameleon Co Ltd., UK, ISBN 0 86022 546 1
[2] Bloch, H.P., Budris, A.R., 2004. Pump User's Handbook: Life Extension. Fairmont Press, Inc Lilburn, Georgia and MARCEL DEKKER, INC. New York and Basel.
[3] ASHRAE, HVAC Applications, Chapters 33, 34, 42, and 48, 2007.
[4] ASHRAE, HVAC Systems and Equipment, Chapters 2, 7, 10, 13, 14, 20, 21, 26, 31, 32, 36, 37, 38, 39, 43, 44, 45, 46, 47, and 50, 2008.
[5] ASHRAE, Refrigeration, Chapters 7 and 41, 2002.

CCHP the Future 10

10.1 Introduction

Combined production of heat and power (CHP) using a single energy source and a prime mover has been recognized as one of the clean energy strategies offered to local governments (EPA 2008). It is evident that improving CHP systems and using them as CCHP systems can magnify their benefits. The best way to imagine the future of CCHP systems is to discuss their real impact on the parameters that are most important for governments and consumers in recent and coming years. In previous chapters, we have investigated the impact of CCHP systems on fuel consumption, pollution reduction, and economic criteria. In the following mutual benefits of CCHP systems for governments and consumers are highlighted from a general point of view.

10.2 Benefits of CCHP for Consumers and Governments

To get a better outlook on the future of the CCHP systems, we should focus on the advantages of CCHP systems for governments and consumers.

10.2.1 Energy Consumption

CCHP systems can reach an overall efficiency of about 75% to 85% depending on the CCHP components and energy demands. This means 25% to 35% fuel savings with respect to SCHP systems, which results in mutual benefits for both consumers and government. This significant reduction of fuel consumption can be translated into less costs in providing, transporting, and distributing (PTD) fuel to the consumers. As a result the local government will pay less for PTD system installation, operation, and maintenance. In addition less fuel consumption means less pollution production and consequently less health care costs. Furthermore, reduction of fuel consumption reduces the fuel bills of consumers.

10.2.2 Omitting T&D Losses and Costs

Generating electricity in the vicinity of the consumer brings important advantages for both governments and consumers. Governments will pay less to develop the electricity transportation and distribution (T&D) grid. It also omits the electricity T&D loss and consequently reduces electricity production, transportation, and distribution costs for governments. It also makes consumers independent from the grid. Well-designed CCHP systems with proper operation and maintenance programs reduce unscheduled shutdowns, which is very important for most consumers. Reducing unscheduled shutdowns increases the reliability of electricity generation, which is very important for companies or institutions whose products and services are very electricity dependant.

10.2.3 Safety

Safety is a critical parameter in production, transportation, and distribution of different types of energy. A centralized, several hundred MW's power plant constructed to provide electricity for a city is very vulnerable to terrorist attacks, war, and natural disasters such as flood, thunderstorms, etc. In an unscheduled shutdown of a centralized power plant all consumers including residential, institutional, industrial, transportation, etc will face the same problem [1–3]. They all are out of electricity. A distributed electricity production system such as a CCHP system is much less vulnerable than centralized electricity production. The reason is very clear since it is actually impossible for terrorists to attack all of the small distributed electricity producers. In addition, in the case of natural disasters many CCHP systems may keep running and emergency help can be diverted to other people who are out of electricity, cooling, or heating. This is very helpful for both governments and consumers. Those who use CCHP systems will most likely have electricity, heating, and cooling in critical situations.

10.2.4 Job Creation

CCHP systems are a new and attractive market that can help governments in producing many new and different job opportunities. CCHP technology development creates jobs at different levels of training, research and development (R&D), consultancy, optimization, manufacturing, marketing, transportation, installation, operation and maintenance, etc. Creating job opportunities is also of mutual benefit to consumers and governments.

10.2.5 Economic Benefits for Consumers

In comparison with SCHP systems, consumers pay a higher initial capital cost with CCHP systems. While when using SCHP systems consumers always pay for operation and maintenance and energy bills, on the other hand SCHP systems produce no income for consumers. On the contrary, CCHP systems make money for consumers through different methods such as fuel saving (lower fuel bills), selling electricity to the grid, and no payments for electricity bills. These positive cash flows can pay back the initial capital cost in a reasonable time. The CCHP system can also be designed to create a positive net present value for the project, which means the project is certainly profitable. Earning money through efficient energy consumption increases the net income of consumers; therefore they are encouraged to manage their energy consumption more efficiently to make the CCHP system more efficient and more beneficial.

10.2.6 Reducing Greenhouse Gas (GHG) and Air Pollution

CCHP systems can reduce CO_2 production up to 50% in comparison with SCHP systems that produce the same amount of cooling, heating, and electricity. Although

CCHP systems can be used in different sectors including residential, commercial, industrial, and transportation, using them even just in the residential and commercial sectors results in a significant reduction of GHG and other air pollution. For example, from the US total energy consumption of 95.02 quadrillion Btu in 2012, residential, commercial, industrial, and transportation sectors consume 20.10, 17.61, 30.54, and 26.77 quadrillion Btu, respectively. This means that the residential and commercial sectors consume about 39.68% of the total energy consumption. The annual energy consumption of residential and commercial sectors for the years 2025 and 2040 based on the annual energy outlook of 2013 is estimated to be 39.20% and 41.07% of the total energy consumption, respectively [4]. According to these statistics and predictions, the residential and commercial sectors have great potential to make use of CCHP benefits, especially in reduction of GHG production and fuel consumption. Decreasing GHG levels decreases the risk of global warming, an international concern. In addition, decreasing any other air pollution such as CO, NO_x, SO_2, etc. improves the air quality and decreases related risks such as disease, acid rains, etc. It is clear that everyone benefits from reduction of air pollution and GHG.

10.2.7 High Potential for Improvement

CCHP systems continue to improve their energy utilization efficiency, they can be further improved by linking them with renewable energy resources and new technologies. CCHP systems can be linked with solar collectors to use solar energy. They can use biofuel as the main fuel, and if available they can use geothermal energy. They can be linked with TES systems to store surplus heat, etc. This high potential for improvement will lead to the development of other new technologies that can be linked with CCHP systems.

10.3 Future of CCHP Systems

In the previous section the mutual benefits of using CCHP systems for governments and consumers were discussed. It is proven that using CCHP systems reduces fuel consumption and air pollution. They can be designed to create economic benefits for the consumer as well. This increases the safety of energy production, transportation, and distribution. It omits many costs and brings profits for governments and consumers. These benefits in addition to government support can help to encourage the use of CCHP systems in newly designed and constructed buildings. Consideration of proper financial support can also encourage consumers to renew their energy conversion equipment and use CCHP systems instead. In addition increasing oil prices due to war or other international instability can encourage the use of more efficient energy conversion equipment such as CCHP systems. Looking at the many advantages of CCHP systems, the future of this new technology looks bright.

References

[1] WADE (www.localpower.org). World Survey of Decentralized Energy. 2004.
[2] Reicher, D. (Texas New Energy Capital). CHP Roadmap Workshop: Five Years into the Challenge. September 2004.
[3] WADE (www.localpower.org). A Lower Cost Policy Response to the North American Blackouts. August 2003.
[4] U.S. Energy Information Administration, Annual Energy Outlook 2014 Early Release Overview.

Symbols

English meaning	Symbol
Apparent extraterrestrial irradiation (W.m^{-2})	A
Absorption chiller	abc
Avoided cost ($)	AC
Collector area (m^2)	A$_{co}$
Solar angular hour (deg)	AH
Analytic hierarchy process	AHP
Ambient	amb
Annual saving	AS
Aggregated thermal demand (kW)	ATD
Azimuth angle (deg)	az
Atmospheric extinction coefficient (-)	B
Boiler	b
Boiler size (kW)	B
Buying electricity	be
Cooling load (kW)	C
Ratio of diffuse radiation on horizontal surface to direct normal irradiation (-)	C
Combined cooling, heating, and power	CCHP
Carbon dioxide emission	CDE
Cash flow (USD)	cf
Combined heating and power	CHP
Closeness number	CN
Collector	co
Coefficient of performance (%)	COP
Main criteria	Cr
Conventional separate production of energy	CSP
Combined weight	CW
Domestic hot water load (kW)	D
Decibel, dry bulb (temperature)	db
Building demand	dem
Diffuse radiation falling on a horizontal surface	dH
Domestic hot water	DHW
Day length (HR)	DL
Direct normal	DN
Diffuse radiation	dθ
Electricity (kW)	E
Electrical	e
Economic	EC
Entropy information method	EIM
Emission production mass (g)	Em
Easiness of maintenance in Iran	EMI
Environmental	EN

English meaning	Symbol
Earnings (USD)	er
Energy utilization factor	EUF
Expenses (USD)	ex
Exergy increase ratio (%)	EXIR
Fuel energy (kW)	F
Fuzzy algorithm	FA
Fan coil unit	FCU
Following electrical load	FEL
Fuel energy saving ratio (%)	FESR
Full load operation	FLO
Following seasonal load	FSL
Following thermal load	FTL
Angle factor between surface and Earth (-)	F_{sg}
Angle factor between the surface and sky (-)	F_{ss}
Electricity grid	g
Natural gas	Gas
Grey incident approach	GIA
Grey incident grade	GIG
Enthalpy ($kj.kg^{-1}$)	h
Heating load (kW)	H
Heat exchanger	HE
Hour	HR
Heat recovery steam generator	HRSG
Heat pump	hp
Pollution index ($kg.MWh^{-1}$)	i
Price index ($USD.kW^{-1}$)	i
Initial capital cost (USD)	I
Thermodynamic state	i
Exergy destruction rate (kW)	i
Internal combustion engine	IC
Import and export limitations	IEL
Input, output	in, out
The domestic hot water system is using the recoverable heat from the engine as well	intg-DHW
Operation and maintenance cost (USD)	I_{OM}
Judgment matrix	J
Lifetime of project (year)	L
Latitude angle of collector position (deg)	LAT
Load ratio	LR
Prime mover options number	m
Mass flow rate ($kg.s^{-1}$)	$\dot{m}$
Maximum, minimum	max, min
Multicriteria decision-making	MCDM
Multicriteria sizing method	MCSM
Miscellaneous	ME
Micro-gas turbine	MGT
Maximum rectangle method	MRM

English meaning	Symbol
Normalized, day number in a year, time period	N
Number of decision-making criteria	n
Nominal (Equipment size)	nom
Number of units in the building	NOU
Net present value (RLS)	NPV
Overall	o
Optimum	opt
Organic Rankine cycle	ORC
Absorber plate of collector	p
Phosphoric acid fuel cell	PAFC
Payback period (year)	PB
Pairwise comparison matrix	PC
Proximity degree	PD
Primary energy consumption (kW)	PEC
Primary energy saving (kW)	PES
Power to heat ratio (-)	PHR
Prime mover	PM
Conventional power plant	pp
Heat (kW)	Q
Useful heat gained by collector $(W.m^{-2})$	q_u
Interest rate of money (%)	r
CO_2 emission ratio	R
Reflectance	re
Recoverable	rec
Relative humidity (%)	RH
Radiation intensity $(W.m^{-2})$	RI
Recoverable and solar	rs
Entropy of information	S
Specific entropy $(kj.kg^{-1}.K^{-1})$	s
Subcriteria	Sc
Selling electricity	se
Separate production	sp
The domestic hot water system is not using the recoverable heat from the engine	sprt-DHW
Stirling engine	STR
Salvage value	SV
Summer, winter	sum, win
Tariff $(USD.kWh^{-1})$	t
Temperature (°C)	T
Thermally activated technology	TAT
Technological	TE
Total horizontal irradiation	tH
Thermal	th
Total solar irradiation	$t\theta$
User friendliness for control and regulations	UFCR
Upward heat loss coefficient, $W.m^{-2}.K^{-1}$	U_L
Volume flow rate $(liter.s^{-1})$	$\dot{V}$

English meaning	Symbol
Weight of each criterion or subcriterion	W
Work, power (kW)	$\dot{W}$
Wet bulb (temperature)	wb
Water heater	wh
Water density (kg.liter^{-1})	ρ_{wtr}
Energy magnitude (power or heat) (kW)	x
Reduction ratio of X (%)	XRR
Number of time increments of a year	Y
Year number	y
Absorptance of plate (-)	α
Effectiveness (%)	ε
Collector-sun azimuth angle (deg)	γ
Solar declination angle (deg)	δ
Thermodynamic first-law efficiency (%)	η
Incident angle (deg)	θ
Linear combination coefficient	λ
Distinguishing coefficient	ξ
Balance coefficient	μ
Exergy efficiency (%)	π
Collector tilt angle (deg)	Σ
Collector azimuth angle (deg)	ψ
Solar longitude (deg)	σ
Transmittance of cover (-)	τ
Exergy rate (kW)	$\dot{\phi}$
Reflectance of the foreground (-)	ρ_g
Thermodynamic dead state (1 atm, 20°C)	0
AND, OR (logical symbols)	$\wedge$, $\vee$
Triangular fuzzy number components	(a,b,c)

Appendix

Appendix 1

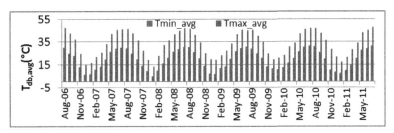

Figure A1.1 Dry bulb temperature over five years for Ahwaz.

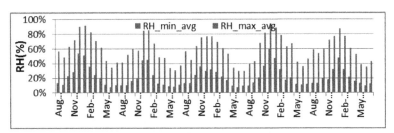

Figure A1.2 Relative humidity over five years for Ahwaz.

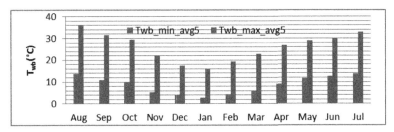

Figure A1.3 T_{wb} of Ahwaz based on five-year average of RH and T_{db}.

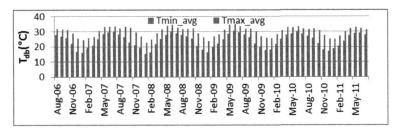

Figure A1.4 Dry bulb temperature over five years for Chabahar.

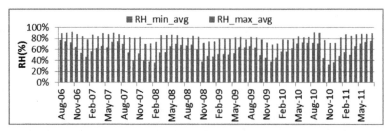

Figure A1.5 Relative humidity over five years for Chabahar.

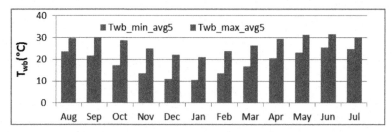

Figure A1.6 T_{wb} of Chabahar based on five-year average of RH and T_{db}.

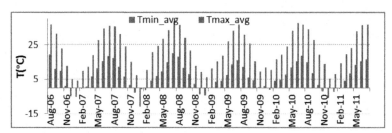

Figure A1.7 Dry bulb temperature over five years for Kamyaran.

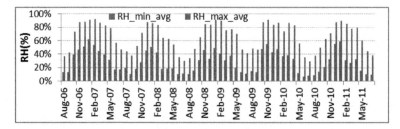

Figure A1.8 Relative humidity over five years for Kamyaran.

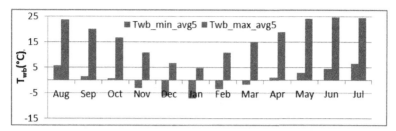

Figure A1.9 T_{wb} of Kamyaran based on five-year average of RH and T_{db}.

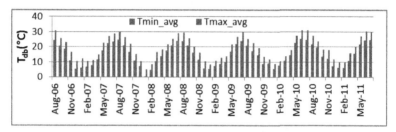

Figure A1.10 Dry bulb temperature over five years for Bandar-Anzali.

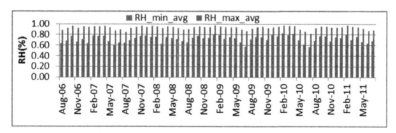

Figure A1.11 Relative humidity over five years for Bandar-Anzali.

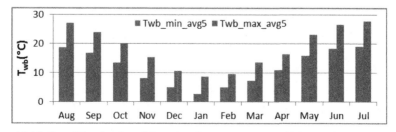

Figure A1.12 T_{wb} of Bandar-Anzali based on five-year average of RH and T_{db}.

Appendix 2

Technical input data to the code:

$\eta_{pp} = 0.35$, $\eta_g = 0.85$, $\eta_b = 0.8$, $\varepsilon_{FCU} = 0.85$, $\varepsilon_{wh} = 0.85$, $\eta_{CHP} = 0.85$, $COP_{abc} = 0.7$, $T_0 = 15°C$, $T_{hg} = 1000°C$, $T_{oil} = 60°C$, $T_{jacketing} = 100°C$, $T_{exhaust} = 540°C$, $Q_{oil} = 0.15Q_{rec}$, $Q_{jacketing} = 0.35Q_{rec}$, $Q_{exhaust} = 0.5Q_{rec}$, $T_{room,sum} = 25°C$, $T_{room,win} = 22°C$, $T_{db,sum,Kerman} = 36.5°C$, $T_{db,win,Kerman} = -2.5°C$, $T_{db,sum,Ahwaz} = 46.74°C$, $T_{db,win,Ahwaz} = 8.73°C$, $T_{db,sum,Bandar\ Anzali} = 29.76°C$, $T_{db,win,Bandar\ Anzali} = 4.61°C$, $T_{db,sum,Chabahar} = 33.55°C$, $T_{db,win,Chabahar} = 16.87°C$, $T_{db,sum,Kmayaran} = 36.82°C$, $T_{db,win,Kmayaran} = -4.88°C$, $LAT_{Kerman} = 30.28°$, $LAT_{Ahwaz} = 31.33°$, $LAT_{Bandar\ Anzali} = 37.46°$, $LAT_{Chabahar} = 25.28°$, $LAT_{Kamyaran} = 34.80°$, $NOU = 8$

Environmental input data:

$i^0_{CO,P} = 4$, $i^0_{CO,C} = 0.1274 / COP_{abc}$, $i^0_{CO,H} = 0.1274$, $i^0_{CO,D} = 0.1274$, $i^0_{CO_2,P} = 700$,

$i^0_{CO_2,C} = 182.04 / COP_{abc}$, $i^0_{CO_2,H} = 182.04$, $i^0_{CO_2,D} = 182.04$, $i^0_{NO_X,P} = 300$,

$i^0_{NO_X,C} = 0.1532 / COP_{abc}$, $i^0_{NO_X,H} = 0.1532$, $i^0_{NO_X,D} = 0.1532$, $i^1_{CO,P} = 0.8$,

$i^1_{CO,C} = 0.1274 / COP_{abc}$, $i^1_{CO,H} = 0.1274$, $i^1_{CO,D} = 0.1274$, $i^1_{CO_2,P} = 430$,

$i^1_{CO_2,C} = 182.04 / COP_{abc}$, $i^1_{CO_2,H} = 182.04$, $i^1_{CO_2,D} = 182.04$, $i^1_{NO_X,P} = 0.7$,

$i^1_{NO_X,C} = 0.1532 / COP_{abc}$, $i^1_{NO_X,H} = 0.1532$, $i^1_{NO_X,D} = 0.1532$

Economic input data (the currency of these data is USD):

$i^0_E = 10.83\ USD$ is the price of each electricity consumption counter.

$$i^1_E = \begin{cases} 1.1(1.534 \times 10^{17} E_{nom}^{-10.33} + 866.4), & 25 \leq E_{nom} \leq 500 \\ 1.1 \times 1420, & 0 < E_{nom} < 25 \end{cases}$$ is the engine cost per kW of

electricity; the coefficient 1.1 is consideration for the heat recovery heat exchangers.

$I^1_{OM} = 1.1(E_{nom} \times 3.3 \times 10^{-5} + 26.5)$ is the operation and maintenance costs.

$i^0_C = i^1_C = (-35.4 \log(C_{nom}) + 431)$ is the chiller cost per kW of cooling. The boiler cost per

kW of capacity is measured as follows:

$$i^0_H = i^1_H = \begin{cases} (1.38 \times 10^9 H_{nom}^{-3.212} + 9.39) & H_{nom} \geq 464.8 \\ (23.54 \times H_{nom}^{-0.08598} - 6.48) & 75.53 \leq H_{nom} < 464.8 \\ 9.74, & 0 < H_{nom} < 75.53 \end{cases}$$

$i_{solar} = 267\ (USD.m^{-2})$ is the plate solar collector cost.

$t_{se} = t_{be} = 0.1188(USD.kWh^{-1})$ and $t_{gas} = 0.377\ (USD.m^{-3})$ are the electricity and gas tariffs.

Printed in the United States
By Bookmasters